BRAND PREMIUM

Also by Nigel Hollis

The Global Brand (2008)

BRAND PREMIUM

How Smart Brands Make More Money

Nigel Hollis
Chief Global Analyst
Millward Brown

BRAND PREMIUM
Copyright © Nigel Hollis, 2013.

All rights reserved.

First published in 2013 by
PALGRAVE MACMILLAN®
in the U.S.—a division of St. Martin's Press LLC,
175 Fifth Avenue, New York, NY 10010.

Where this book is distributed in the UK, Europe and the rest of the world,
this is by Palgrave Macmillan, a division of Macmillan Publishers Limited,
registered in England, company number 785998, of Houndmills,
Basingstoke, Hampshire RG21 6XS.

Palgrave Macmillan is the global academic imprint of the above companies
and has companies and representatives throughout the world.

Palgrave® and Macmillan® are registered trademarks in the United States,
the United Kingdom, Europe and other countries.

ISBN: 978–1–137–27991–0

The Library of Congress has catalogued the hardcover edition as follows:

Hollis, Nigel.
    Brand premium : how smart brands make more money /
Nigel Hollis, Chief Global Analyst, Millward Brown.
        pages cm
        ISBN 978–1–137–27991–0 (alk. paper)
        1. Brand name products. 2. Branding (Marketing) 3. Product
    management. I. Title.

HD69.B7H6463 2013
658.8'27—dc23   2013008881

A catalogue record of the book is available from the British Library.

Design by Newgen Knowledge Works (P) Ltd., Chennai, India.

First edition: October 2013

10 9 8 7 6 5 4 3 2

Printed in the United States of America.

Dedicated to Gordon Brown

Contents

Foreword

I can trace my digital footsteps back to a flurry of emails between Nigel and me in March 2010, when this whole adventure kicked off. We had decided that we should do more to share our wider learning about brands with our clients. In particular, we wanted to use this broader knowledge to tackle the specific challenges and opportunities of individual brands. We have always believed that we should be helping businesses, not just through the application of findings from a single piece of research, but by harnessing all of Millward Brown's accumulated learning. Over 40 years, we have collected data in over 70 markets and across thousands of brands; we discuss brands with marketing people and work with their agencies every day of the year. More than this, we tie what we know to financial success in the market.

Back in 2010, there was an established view within the company about how brands worked, but it hadn't been revisited for some time. We felt that if we were going to take our brand knowledge out to clients, then we should be taking them the very latest insights. So we began reexamining all of our datasets, talking to colleagues inside and outside the company, and then refining and reworking the models that we had. There was no talk about writing a book back then—or even about influencing the kind of measures we use on a day-to-day basis; instead, we simply aimed to pull together our best understanding of how brands work in order to help our clients. Six months later, we had our model and began road testing it. That meant presenting it at conferences, using it as a framework to analyze real brand predicaments, and talking about it to clients whenever we could. A year later, by the end of 2011, the model had developed considerably. It had become a bit more complex and a lot more robust. The idea of building a more formal workshop that could be run by senior people in the company started to form at that point, as did the idea of writing a book.

Nigel's first book, *The Global Brand*, was published to great acclaim in 2008, so enough water had flowed under the bridge for the idea of writing on this scale to be appealing once more. When you know you're going to write something, you become a hoarder of articles, quotes, ideas, and snippets of conversation in the hope that you can weave all these elements into a wonderful tapestry. So Nigel had another year of running workshops, talking at conferences, discussing ideas with clients, and generally collecting the anecdotal material and insights needed to help add richness and texture to the story.

This book is particularly powerful because it operates on three levels. First of all, it's based on robust analysis from some of the biggest databases in the world, populated with data on consumer loyalty and how it is affected by communications and other activity of all kinds. Second, it provides a layer of anecdotes and examples culled from the two years during which this model was incubated and developed. Third, it offers Nigel's own passionate take on the importance of brands, uncovering and explaining all the levers that can be pushed and pulled to make a brand financially successful. He cares not only about brands being successful but also about marketers being successful; he wants the discipline of marketing to be recognized for what it can contribute to business success. A subtext of the book is the need for marketers to become more central to the running of businesses, because they are fulfilling one of the most vital roles in wealth creation. We have encountered too many examples (even in some of the world's biggest companies) in which the brand assets of the company are poorly understood; these brands would benefit hugely from the kind of thinking captured in these pages.

Many marketing books take one particular aspect of marketing success and elaborate on it. This book is different; it provides a dynamic model that describes all the influences that are (or should be) under the control of the marketing team. It lays out an overall plan, a checklist that any brand owner can use to understand how well his or her brand is performing on each dimension that leads to financial growth.

We are at the end of one adventure but the beginning of another. All of the thinking in this book has led us to reflect upon the metrics we use in all our work, on the questions we ask, and on the analyses we conduct to provide recommendations to brand owners. In a way, this reflection was the ultimate test of the ideas in this book. We devised new approaches based on this thinking that would only have had merit if they more accurately predicted and explained current market share and future share growth. Nine months of intensive research and development explored whether these ideas could concretely be tied to financial outcomes. The fact is, they can. They even provide a method for understanding how strong brands can command a price premium, and that is the ultimate validation of these ideas.

Now we have the excitement of knowing that the ideas Nigel captures in this book are both practical and powerful; this book is a state-of-the-art guide for marketers to grow their company's financial success.

—GORDON PINCOTT,
Chairman, Global Solutions, Millward Brown

Acknowledgments

First, my grateful thanks to Gordon Pincott for his part in developing the ValueDrivers framework, his counsel on the book, and being a good travel companion.

Second, thanks to Jill Davies for her patience while I was writing the book and for her valuable feedback on the final draft.

Third, I would like to extend my thanks to the following people, who kindly agreed to be interviewed for this book and shared their expertise and observations with me.

Alina Asiminei, marketing director, Philips, Brazil

Duncan Blake, brand director, BP PLC, UK

Tom Buday, head of marketing and consumer communication, Nestlé, Switzerland

José Calderoni, marketing and sales manager, Volaris, Mexico

Juan Carlos Canales, marketing director, Casa Cuervo, Mexico

Paula Castellan, marketing director, Santa Clara Cafés, Brazil

Gary Chu, CEO, General Mills China

Jeremy Diamond, partner/chief strategy officer, Distillery, USA

Luciana Fortuna, marketing director, PepsiCo, Brazil

Daniel Keller, director of brand strategy, Orange Group, UK

Barbara Lamprecht, director of brand management and strategy, Skoda China

Rosana Lourenço, marketing director, Hanes Brands, Brazil

Patricia Martínez, PR and corporate communications director, Danone, Mexico

Leandro Medeiros, marketing director, Diageo, Brazil

Andrea Mello, marketing director, Mondelèz, Brazil

Mark Murray, global consumer planning director, Diageo, UK

Renata Petrovic, marketing director, HSBC, Brazil

Bettina Walker, marketing director, Johnson & Johnson, Brazil

Zhenzi Liu, senior assistant for general manager of CMBC Branding Sector, China Minsheng Banking Corp., Ltd.

In addition, I would like to thank my many colleagues who contributed ideas, analysis, and case studies for the book. A special shout-out goes to Helen Fearn, who conducted much of the analysis that led to the ValueDrivers model and subsequent analysis for the book. I cannot list everyone else involved here, but specific thanks go to Jorge Alagon, Fernando Alvarez, Jackie Bartolotta, Dale Beaton, David Chantrey, Elspeth Cheung, Nick Cooper, Monica Garcia, Enrique Gonzalez Sainz, Miquet Humphryes, Graham Page, Silvia Quintanilha, Josh Samuel, Albert Sim, Mario Simon, Duncan Southgate, Scott Spence, Jason Spencer, Dominic Twose, Sarah Walker, and Doreen Wang.

And last but not least, I would like to thank Keri Cole for her invaluable help in editing my ramblings into something coherent.

Preface

Is marketing dead?

In August 2012 Bill Lee, president of the Lee Consulting Group, published a post on the *Harvard Business Review* blog under the provocative title, "Marketing Is Dead."[1] He states: "Traditional marketing—including advertising, public relations, branding and corporate communications—is dead. Many people in traditional marketing roles and organizations may not realize they're operating within a dead paradigm. But they are. The evidence is clear."

I totally disagree with this statement. I believe the practice of marketing can still generate significant financial value, but many CEOs do not believe that to be the case. Lee references a study from the Fournaise Marketing Group that finds that 73 percent of CEOs think marketers lack business credibility and fail to drive financial growth. Unfortunately, findings like these are far from isolated.

But I am not convinced that this study's results really indicate that CEOs doubt the value of marketing; rather, I believe CEOs doubt the value of the way that their marketers practice marketing.

The issues cited by CEOs in surveys and interviews fall into three main categories:

1. Marketers focus on brand, brand equity, and brand positioning without linking these concepts to the outcome for which CEOs are held accountable: financial value growth.
2. Rather than prove that their spending is an investment that drives business results, marketers focus on efficiency and cost cutting to prove they are creating value.
3. Marketers fail to offer a coherent rationale as to how the practice of marketing leads to positive consumer behavior and financial outcomes such as stronger demand, higher prices, and stronger growth.

This book aims to address these three areas. I will provide evidence that demonstrates how marketing is capable of driving financial value growth without merely cutting costs and maximizing efficiency. In addition, I will document how the things that marketers believe are important help generate that growth.

PART I

WHAT'S IT ALL ABOUT?

I have now worked in the domain of brand building for three decades.

On one hand, I find that statement incredibly depressing. Where did the time go?

On the other hand, over the years I have had the chance to make some important observations about what it takes to grow and maintain a successful and profitable brand. As a market researcher by profession, I have had the opportunity to study brands and advise marketers in many different product categories and countries. Rather than getting lost in the day-to-day practice of brand management, I have had the advantage of being able to stand back and observe the common patterns of success and failure.

Three general observations underpin my perspective on what it takes to build a growing, profitable brand today.

The first is that although our world is changing quickly, what matters to us as human beings—the things make us willing to pay for a brand—is not. We are still drawn to things that add meaning to our lives, and we are willing to pay more for them.

Second, effective brand building is most important when the other variables in the mix—the sourcing of raw materials, manufacturing, distribution, and the like—no longer yield significant competitive advantage.

Last, but not least, marketers must be able to keep their ultimate objective in mind without being overwhelmed by day-to-day implementation. Without a clear idea of what makes your brand meaningfully different from the competition, time and money will be wasted on dead-end initiatives that have little hope of payback.

BRANDS OUTLIVE COMPANIES

In 1979 I joined the company then known as Cadbury Schweppes and worked in Cadbury Foods Ltd. The importance of brands was a central tenet of the company. No one questioned the value of brand building, and it was an accepted fact that if a product was worth producing, then it was worth investing the time and resources required to build a brand around it. The evidence for this belief rested in the portfolio of strong brands that Cadbury Schweppes owned: Cadbury's Dairy Milk chocolate, Flake, Crunchie, Marvel milk powder, Smash instant mashed potato, Schweppes Tonic, Sunkist, and Canada Dry.

The Cadbury Schweppes for which I worked no longer exists. The foods division was subject to a management buyout in 1985, Cadbury Schweppes de-merged in 2008, and Cadbury was bought by Kraft in 2010. Factories that I once visited as an eager trainee have been decommissioned or sold. Colleagues from the time have moved on to new ventures and then on again. But the brands still exist. Indeed, the brands are sometimes the only assets that survive the vagaries of changing ownership.

BRANDS OFFER THE POTENTIAL TO GROW FINANCIAL VALUE

When Kraft acquired Cadbury for $18.9 billion, it did so to gain control of the company's portfolio of valuable brands.[1] At the time of the acquisition, I estimate that Cadbury's Dairy Milk accounted for over 15 percent of Cadbury's sales, making it worth more than $1.6 billion, but Kraft was not buying the brand for its value in 2010. Instead, Kraft was betting on the ability of this iconic brand to produce a continued and growing profit stream for years to come.

As bets go, this one does not seem that risky. Cadbury's Dairy Milk has a 108-year track record that has seen the extension of the brand's appeal around the world, and it provides Kraft with a strong foothold in fast-growing markets like India, China, and Brazil. Two years after the acquisition, Cadbury India enjoyed the best performance in the history of the company with sales growing 40 percent, in part because of the rapidly growing popularity of Cadbury Dairy Milk.[2] Growth in India alone would not justify the $11.5 billion Kraft borrowed to finance the takeover, but if it is indicative of performance in other developing markets, then the brand promises significant earnings potential to its new owner.

THAT WAS THEN, THIS IS NOW

A marketer's life today is far more complex than it was when I started work. Just think of all the things that have changed over the last 30 years in the area of advertising alone.

We have gone from a time when TV was clearly the dominant medium, reach was cheap, the Internet was science fiction, word of mouth was ignored, video was synonymous with TV, you bought your media, and social networking was an evening down at the pub with the brand team.

Now we have a dramatically different media scene. TV may still be the top dog, but it is looking pretty chewed up as other media are all trying to take bites out of its budget, and that fragmentation means mass reach is far more expensive. The Internet is a global reality that enables instant sharing. Word of mouth marketing is now visible and sexy. Video is now on our computers and smartphones, on billboards, and in taxis and elevators, and viral video is the latest free lunch. If people like something, technology enables them to act as a network of their own and pass it on to their friends at no charge, and social media like Facebook and Twitter mean that people can share stuff with friends around the world.

From my point of view, all that these changes affect is how we deliver our marketing content. It is undeniable that the media world has changed, but has the content that people appreciate changed? Is what makes people want to share a video with their friends really different from what made an ad memorable three decades ago? Based on the evidence I have found, the answer is no. In order to engage attention and be memorable, people must find an ad to be enjoyable, interesting, and involving. Those same criteria are necessary if people are to share an ad with their friends, but in addition, they need to find the content funny, edgy, gripping, sexy, and different from the content of other ads. I believe we are playing exactly the same game as before, but the bar has been raised to new heights.

And what about brands? Has what people expect of a brand changed? I would argue that it has changed less in nature than in degree. People's expectation of quality is far higher than it used to be, but people still want brands to solve problems, empower them, and make them feel good. The importance of brands has morphed from being a signal of quality to being a signal of self-worth. However, people still seek out the brands that they feel are right for them. They are still willing to pay a premium for a brand that they believe meets their needs better than others do. They are intolerant of poor experiences and inept marketing tactics that do not respect their needs, time, and intelligence. Is this new? No. But the ability for people to signal their discontent through Twitter, blogs, and consumer review sites is new. Thanks to the Internet, positive and negative brand experiences are shared more widely now than ever before.

◼◼◼ NEW WORLD, OLD MOTIVATIONS

Technology is facilitating old behaviors far more than it is creating new ones. People have always searched for information and advice when making a purchase. In the past, they may have relied on input from friends or family. Now, they simply use Google to access a host of free information and advice. People have always price shopped. In the past, they might have visited several stores in order to check out the alternatives. Now, they can get most of the work done during their morning commute using their smartphone. People have always talked about brands. In the past, they did so around the kitchen table or over the garden fence. Now, they do so on Facebook with friends around the world. People are doing what they have always done, but they are doing it far more easily, quickly, and widely.

While these changes pose problems for marketers, they also offer new opportunities. Savvy marketers know that the motivations for building brands today are the same as they were 30 years ago, but now marketers have new tools that allow them to communicate and interact with customers and prospects on a far more informed basis. But with so many choices available, how do marketers persuade people to choose the brand they are promoting? Marketers just have to remember that unless they are offering something meaningful and different, people will resort to what they have always done, which is to look for the best deal.

◼◼◼ THE NATURE OF GOOD MARKETING REMAINS UNCHANGED

The fundamental nature of good marketing remains unchanged: to create brands that people find valuable and, by doing so, create value for the company that owns the brand. This task is far more than just "advertising." We still need to identify and satisfy people's needs and motivations. At the most basic level, this involves identifying opportunities to satisfy specific needs better than the alternatives can or to do so in a new and different way. In this respect, marketing should be responsible for shaping a company's innovation agenda, ensuring that there is a good match between what the company delivers to the market and what people find useful and worth paying for.

Building on this role, marketers need to identify how best to communicate with their target audience. They must understand how the brand will fit into people's lives and then position it to add intangible value to the intrinsic value of the product or service. To do so, marketers need to gain an intimate understanding of their audience—to put themselves in consumers' shoes, as it were. They need to know what people want to experience in life and how brands fit with those desires. They need to know what cues trigger people to behave in certain ways and make different choices. They need to anticipate what will change and what will stay the same in this ever-accelerating world so that they can evolve their approach alongside changing consumer behavior and attitudes.

JUSTIFY YOUR PREMIUM PRICE

So, yes, the world of marketing is infinitely more complex than it was when I started my career over three decades ago; that much is indisputable. But the essential process by which marketing builds a brand and adds value to a business has not changed. Why? Because human nature has not changed. And that means the potential to build strong, valuable brands is as great now as it was then—perhaps even more so.

In spite of this continuity, the value of marketing as a practice is under greater scrutiny today than ever before. Marketers are constantly asked to prove their return on investment and to do more with less. My own observations suggest that instead of rising to the challenge, the marketing profession is shooting itself in the foot. At a time when brands are more valuable than ever, bought and sold for many times their annual revenues, we are losing sight of what makes brands enduring, valuable assets. We ignore what makes people want to buy brands and be willing to pay a premium for them. Why? Because so many day-to-day tasks demand our attention. We are so busy executing that we have forgotten why we are doing what we are doing, and we rely on metrics to guide our actions without judging their relevance or utility.

Consumer motivations have not changed, and neither have the ways that brands make money. There are five basic ways to create more value from a brand:

1. Encourage more people to buy the brand
2. Encourage people to buy the brand at a price higher than the alternatives
3. Encourage people to keep buying the brand
4. Encourage people to buy the same brand but for new occasions or in new categories
5. Do all of the above, but more efficiently

Unfortunately, many marketers and CEOs appear to be fixated on the first and last of these strategies to the exception of the middle three, particularly justifying a price premium. The ultimate role of a strong brand is to command a price premium over comparable products. All too often I observe brands chasing additional volume at the expense of their price premium and future profit stream. It is debatable whether such tactics pay off in the short term, and all the evidence suggests that they undermine long-term value. Again, what is brand building about if not creating sustainable financial value, a reoccurring profit stream over years, not just months?

▰▰▰ A ROADMAP TO BUILDING FINANCIAL BRAND VALUE

That is where this book comes in—to provide a clearly documented roadmap to make sure that your brand is adding sustainable financial value to your business.

The roadmap is based on a conceptual framework called ValueDrivers, jointly developed by Gordon Pincott, chairman of global solutions at Millward Brown, and myself. Our framework for building brand value is not informed by experience alone. We have been fortunate to be able to draw on all the resources at Millward Brown's disposal, including access to some of the best marketers in the world today; insight from colleagues with specialist knowledge in neuroscience, brand equity research, and brand valuation; and analysis of the world's largest brand equity database, BrandZ™. Since 1998, Millward Brown has interviewed people around the world about their attitudes toward brands, and the database now includes data on over ten thousand brands from over two hundred product categories and over 40 different countries.

In addition to these company resources, I have drawn on a number of other sources of information. Most important among these are our clients. Gordon and I have discussed the framework detailed in this book with some of the most experienced marketers in the world. We have used the ValueDrivers workshop to explore specific brand issues in-depth for a wide variety of international clients, and in preparation for writing this book, I interviewed senior marketers in China, Brazil, Mexico, the United Kingdom, Switzerland, and the United States.

Last, but not least, in order to illustrate specific ways in which brands have created demonstrable value, I have drawn on case studies from the Institute of Practitioners in Advertising (IPA) and winners of the Effie Awards sourced from the invaluable Warc knowledge bank. Euromonitor has also proved a useful source of information on trends and market share data.

By drawing on all of these resources, I aim not only to illustrate what makes a strong brand but also to document its impact on the bottom line. In the first third of this book, I will lay out our general knowledge about brands, specifically why and how they create value for consumer and brand owner alike. In the remainder of the book, I will condense that learning into a set of guidelines for generating financial value growth from any brand.

Chapter 1

How Marketing Adds Financial Value to a Business

Since 2006, Millward Brown has created an annual ranking of the world's most valuable brands on behalf of its parent company, WPP. The BrandZ Top 100 Most Valuable Global Brands Ranking is different from other rankings because it integrates financial data with attitudinal survey data to identify the degree to which consumers' beliefs about brands drive their purchase decisions. We call the attitudinal component the *brand contribution*. The higher the brand contribution, the greater the importance of the brand in driving financial value compared to other tangible and intangible business assets.

To create the ranking, Millward Brown's consulting unit, Millward Brown Optimor, combines publicly available financial data, sales data from Kantar Worldpanel, and results from 150,000 consumer interviews. Table 1.1 shows the top ten most valuable brands in the 2013 ranking.

These top ten brands have increased in value by nearly $500 billion since the first ranking. Some of the growth, but not all, reflects the rise of leading technology brands over that period. Against the technology tide of Apple, Google, and IBM, brands like Coca-Cola, McDonald's, and Marlboro have not only held their own but grown. McDonald's was not ranked in the top ten in 2006, but strong growth over the intervening period means it is now ranked fourth.

The brand contribution varies depending on the brand and the product category. Coca-Cola is far more dependent on consumers' desire to buy its product than is a brand like GE (ranked eleventh, with a brand contribution index of 2), for which purchase decisions are heavily influenced by business requirements, corporate buying policies, budgets, and so on. A high brand contribution is most often found in categories such as luxury goods, for which the emotional connection to the consumer is more salient than any functional benefit. While brand contribution differs across categories, within

Table 1.1 BrandZ Top 10 Most Valuable Brands 2013

Rank 2013	Category	Brand	Brand Value 2013 ($M)	Brand Contribution Index	% Brand Value Change 2013 vs 2012	Rank Change
1	Technology	Apple	185,071	4	1%	0
2	Technology	Google	113,669	3	5%	1
3	Technology	IBM	112,536	3	-3%	-1
4	Fast food	McDonald's	90,256	4	5%	0
5	Soft Drinks	Coca-Cola	78,415	5	6%	1
6	Telecoms	AT&T	75,507	5	10%	2
7	Technology	Microsoft	69,814	3	-9%	-2
8	Tobacco	Marlboro	69,383	3	-6%	-1
9	Credit Card	Visa	56,060	4	46%	6
10	Telecoms	China Mobile	55,368	3	18%	0

a product category, better-known and better-liked brands create stronger demand for that brand's product or service, are worth more, and tend to grow revenues and profits faster than their peers.

Brands are valuable not just for the profits they generate today but also for their future earnings. A brand's current momentum is important to its future earnings potential. Momentum takes into account product category, country growth rates, and consumers' attitudes toward the brand. Technology and telecom brands typically have high momentum scores because of strong momentum takes category growth rates and booming consumer demand in developing economies, but strong consumer demand helps give brands like McDonald's strong growth potential.

████ SMART BRANDS OUTPERFORM THE S&P 500

The Standard & Poor's 500, or S&P 500, is a stock market index based on the market capitalizations of five hundred leading companies publicly traded in the U.S. stock market. In terms of stock prices, the strong brands measured in the BrandZ ranking consistently outperform their peers and the S&P 500.

Strong brands, defined by the perceptions of their target consumer, not only create more business value but also command higher share prices as a result. Figure 1.1 compares the share price performance of companies that are strong brands in our BrandZ portfolio with the S&P 500 and shows that over time, strong brands have significantly outperformed the average, showing a 28 percent improvement in share price over the S&P 500 since

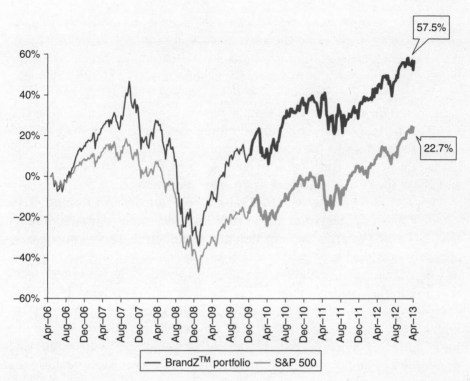

Figure 1.1 BrandZ™ Strong Brands Portfolio vs. S&P 500 (April 2006–December 2012)

Source: Bloomberg, MB Optimor analysis.

April 2006 in spite of the impact of the Great Recession brought on by the 2007 financial crisis.

What is noteworthy is that while the average share price of these strong brands was equally affected by the advent of the recession, it bounced back faster, regaining the 2006 level at least a year earlier than the overall index. Strong brands are resilient in times of crisis and recover more quickly than others, in part because they are important to and trusted by their consumers.

SO WHAT IS A BRAND?

In my previous book, *The Global Brand*, I defined a brand as "a set of enduring and shared perceptions in the minds of consumers. The stronger, more coherent and motivating those perceptions are, the more likely they will be to influence purchase decisions and add value to a business."[1]

People's perceptions of a brand are rooted in their ideas, feelings, impressions, and experiences, and the job of marketing is to help ensure that people's perceptions are as motivating as possible; to make people value the brand more than the alternatives; and, in some cases, to blind them to the fact that there are acceptable alternatives. Marketing is all about creating value based on consumers' intangible memories and associations. The end result of successful marketing is that more people desire a product and are willing to pay the price asked for it—a price that will be more than the product is intrinsically worth because consumers value some aspect of it over and above the tangible product. They value the brand.

People value brands because brands are shortcuts to meaning. That meaning may originate from the function the product performs and a belief that the brand performs that function in a meaningfully better way than do alternatives. More likely, people intuitively recognize that the product does its job well but is not functionally different from the alternatives. Instead, they think the brand is more interesting, likeable, inspiring, aspirational, compelling, or shareable. Ultimately, people think the brand is more meaningful, and people value things that provide them with rewarding and meaningful experiences—brands are no exception. We will explore the dimensions of brand meaning in more detail in Chapter 3, particularly the importance of being meaningfully different from competitors.

A FAIR EXCHANGE OF VALUE

To create sustainable value, a transaction must be viewed as a fair exchange by both seller and buyer. If a brand is to survive and prosper, then its consumers must believe that each purchase is a fair exchange: the service rendered must justify the price paid. Repeated transactions, the ultimate goal of any strong brand, require the consumer to feel that he or she is continuing to enjoy fair value each time he or she uses the brand. Repeated positive experiences are the foundation of long-term brand value (even if the experience is not consciously appreciated every time). The event most likely to destroy brand value is the unexpected bad experience, one that forces the customer to consciously reconsider his or her feelings about the brand.

Product experience may be the cornerstone of brand experience, but it is far from the totality. Brand experience encompasses everything from the earliest fleeting impression of a brand to the latest interaction; from the positive feeling created by seeing a funny TV ad to the discussion of the brand's shortcomings with a friend. It is the job of marketing to shape those experiences to the best advantage—to shape expectations, frame experiences, play down weaknesses, and keep positive associations alive in consumers' minds. Far from letting go in the face of growing consumer control, marketers must work even harder to create a positive and meaningful brand experience. In order to manage our brands effectively, we must first have a clear perspective on what really matters to our consumers. Then, and only then, can we tailor the total brand experience to best effect.

EVERYTHING IS BRAND

When does the taste of the product stop being a function of the recipe and start being a memory of a pleasant experience? There are two sides to any brand: the tangible assets that the brand owner controls and the intangible assets that arise from consumers' feelings about the brand. Because marketers usually concern themselves with the intangible—the ideas and associations that they promote with the intention of making the brand more desirable—they risk decoupling those ideas and associations from the tangible brand experience. That risk is exacerbated by the growing focus on digital communication, since ideas and information are paramount in the virtual domain.

In reality, the tangible assets have far more influence than the "advertising" on whether someone buys a brand and sticks with it. Even if advertising alone leads someone to consider purchasing a brand, most buyers will try to check it out before buying; they will try to anticipate what the brand experience will be like. If marketing is to be really effective, then it needs to encompass both the tangible and the intangible. Sometimes the biggest return on marketing investment will come from simply highlighting a good product experience and the feelings it evokes.

In a recent post on the *Harvard Business Review* blog, Dan Pallotta drives home the point that a brand is a two-sided coin. Pallotta proposes that "brand is everything, and everything is brand."[2] Your brand is your strategy, call to action, customer service, communication with customers, and, yes, your logo and visuals too. He concludes, "Ultimately, brand is about caring about your business at every level and in every detail, from the big things like mission and vision, to your people, your customers, and every interaction anyone is ever going to have with you, no matter how small."

To Dan, a brand is essentially a promise incarnate, but there is another side to this coin: how well the brand lives up to its promise in the mind of the customer. This is what will determine how much the brand is actually worth to someone. You can care about your business as much as you like, but unless your customers care too, your brand is not worth anything. To create value from your brand, you must understand what matters most to your customer.[3]

While a brand is everything its owner controls, it is also everything it suggests in the minds of its customers. A brand is the ideas, the memories, and the feelings evoked every time someone thinks of the brand. To create value, those mental associations must make the product or service more salient, more interesting, or more compelling than the alternatives. They must make the brand meaningfully different in some way. That difference does not need to be tangible or significant, but to create value for the brand owner, that difference must resonate with the potential customer more strongly than competitive brands do. The degree of differentiation required for a brand to create value will depend on the nature of the brand and category. The key question to ask is whether your brand is perceived to be different enough given its competitive context.

So Dan Pallotta is right—"everything is brand." The person on the street makes no distinction between marketing, production, sales, or customer service—those distinctions are reserved for the business world alone. To the individual consumer, these are simply different aspects of the same brand. A compelling online video is just as much a part of a consumer's brand experience as is using the product, seeing the brand on sale in the store, or talking to a customer service representative on the phone. These interactions are all part of the brand experience, and everything a brand does should be aligned to make that experience meaningful. The objective should be to build, reinforce, and enhance the ideas and associations that resonate with customers and drive their predisposition to buy the brand and pay a premium for it. The end result will be the creation of value for both brand owner and brand buyer. And if consumers believe they are getting good value from a brand, then they will talk about it. They will become advocates for the brand.

OVERSTATED EXPECTATIONS— DON'T BE DELUSIONAL

It is the dirty little secret of marketing that much of the time its promises of growth are grossly overstated.

I can understand why this happens, because I have experienced it firsthand. When I ran a small Internet start-up for Millward Brown just before the dotcom bust of the late nineties, I was responsible for both management and marketing. I submitted several budgets with expectations of significant revenue growth, only to see those expectations dashed. The problem was not that we weren't publicizing the company enough—Millward Brown Interactive had great publicity, more coverage than even Millward Brown—but we just could not hit the sales numbers. In preparing my budgets, I overestimated the demand for what we had to offer. Money was flowing freely, and there was little need for market research to justify whether online advertising worked or not. The stuff that we could do, which was traditional market research conducted online, did not pay the bills. And the radical new approach—the vision on which the company was founded—proved difficult to build and implement. (As an aside, that is why we ended up buying Dynamic Logic some years later. They shared the vision and got the business model right. If you can't beat them, buy them.)

I think marketing's promise gets exaggerated for a number of reasons.

First, marketers know that they will be judged by how well they drive top-line growth and try to act accordingly. However, the relationship between what marketers do and the effect of those actions on financial value is not direct—inputs do not translate immediately or directly to outputs but are instead mediated by consumers' mindset and behavior—and cannot easily be separated from the state of the business itself. A brand is everything that the company does, not just its marketing, and marketing itself is a multiplier of what the company does—it is not independent. If the business is flagging, then so too will the return on its marketing investment.

Rory Sutherland, vice chairman of Ogilvy Group, suggests that marketers have become their own worst enemies by trying to justify every action with a spreadsheet. In a YouTube video he states, "What marketers have developed is a sort of Stockholm Syndrome where we have been so beaten up by procurement and finance and other people that we have started to take on the worst qualities of our abusers."[4]

He concedes that justifying marketing actions with numbers is well and good, but he suggests that all too often we predict linear outcomes that scarcely pay for the investment, when the magic of marketing lies in the fact that relatively small actions can sometimes have a very large payoff. He states, "The glorious thing about marketing is that a very small intervention can have huge effects."

Second, expectations of change are often simply out of alignment with reality. Millward Brown Interactive had enough trouble hitting its sales targets at a time when the money put into the Internet was going through the roof, and the biggest drivers of business growth are economic demand and category demand, not market share gain. So, for example, between 2002 and 2007, simply holding market share constant in the automotive market in China was going to guarantee massive returns. Unfortunately, most marketers are operating in categories in which the growth rate is limited. Their challenge is to grow market share, and, unfortunately for the optimists in marketing, market share proves remarkably difficult to change.

Finally, working on a quarterly or annual planning horizon makes it easy to get duped into thinking that the current planning cycle is the only one that counts. However, if you look across several years, then you get a different picture. Successful innovation works on a very different beat than the rest of the business cycle. Groundbreaking innovation only comes once in a while, and, even if effective, new marketing campaigns take time to impact all the potential buyers of a category. Binet and Field find that positive pricing effects—one of the most overlooked benefits of marketing investment—are most demonstrable over a three-year time frame.[5] Competitive advantage gained by brand building must be viewed in the context of years, not months.

THREE WAYS IN WHICH MARKETING DRIVES BRAND VALUE

With the preceding caveats in mind, let me say for the record that marketing can drive enormous financial value growth, particularly when it is aligned with a great business model, and driving top-line growth is really just the tip of the iceberg. The contribution of marketing is far more than just driving the acquisition of new customers; successful marketing also helps make sure you keep the ones you have. In the next section of this chapter, I will consider three ways in which marketing adds value to a business.

1. Driving Financial Growth: Amplifying Tangible Advantages

Marketing drives financial growth best when it is aligned with meaningful and differentiating product offers. Look back across the history of any category. Substantial market share changes happen when a brand introduces something new and meaningfully different to the category. Sometimes the innovation will transcend an existing category or create an entirely new one. For example, the launch of Nespresso has fundamentally changed the way that many people in Europe make coffee at home. The 2011 Nestlé annual report states that Nespresso's revenues grew 20 percent over the course of the year, to well in excess of $3.3 billion. By contrast, the iPhone did not create a new product category; it simply reset the gold standard of what a phone could be. Product-based innovation that creates a tangible advantage over the existing alternatives causes people to stop and reconsider their existing choices, giving new users a compelling reason to choose. This is when marketing can achieve the most effect and when it is worth investing more than normal to ensure that people hear about and understand the new offer.

Innovations like these do not come along often. In established consumer packaged goods categories, the gap between real innovations can be measured in decades. Even in the fast-paced world of technology, there is usually a gap of a couple of years between innovations like the Razr, BlackBerry, and iPhone. When a company does create an innovation, marketing is responsible for spreading the word and letting people know what the brand represents. If you know what aspects of the total brand experience make it valuable to people, then this task is relatively simple. If you don't, then it can seem daunting and all-consuming.

When a brand possesses an intrinsic product or service advantage, then marketing simply needs to enhance perceptions of the value that is already there and amplify that meaningful experience out into the marketplace. Marketing puts relevant and differentiating elements of the product experience under a spotlight. This helps focus people's attention on the good aspects of the experience and provides a point of comparison that favors the advertised brand. However, competitive advantages like these are often short-lived, and, as we shall see, the critical challenge for marketing is not only to create those initial perceptions of difference but also to maintain the feeling of difference for as long as possible.

2. Creating Financial Growth through Intangible Differentiation

In the absence of radical product innovation, marketing can create a similar effect by finding ways to get people to re-envisage what the brand stands for. The innovation is extrinsic, not intrinsic. In cases like these, marketing creates additional and related experiences that improve the brand's value to its customers.

This strategy may take the form of a repositioning exercise that finds a new way to make what the brand stands for different and relevant to its potential consumers. A brand can create perceptions of difference based on its values, personality, tone of voice, and the causes it espouses. But—and this is a big but—this form of advantage is only as good as your last execution. Relying solely on intangible differentiation puts the marketer on a treadmill that requires continued successful implementation.

That said, even a generic category benefit can create a point of difference if a brand makes the benefit its own. Take, for instance, Johnnie Walker and the idea of progress. After a two-year research program designed to identify a motivating idea on which to base its marketing, the brand team identified the concept of progress as one that resonated with consumers around the world. Rather than being satisfied with demonstrating their existing status through the traditional trappings of success, the research found that men now aspired to demonstrate continued personal development and achievement.

Mark Murray, director of global consumer planning and brand building DWBB at Diageo, states: "Any brand of whiskey could have claimed the idea of progress, but Johnnie Walker got in first with an iconic campaign and claimed that territory."[6]

By identifying itself with the universal values associated with the concept of progress and allying that idea with the image of the striding man and a strong marketing campaign, Johnnie Walker managed to carve out a "different enough" positioning in people's minds. The end result has been a dramatic reversal of fortunes for the brand, with strong increases in sales and market share.

To the best of my knowledge, Johnnie Walker did not change its product; the brand relied on marketing to create a perception of difference. But that did not stop the "Keep Walking" campaign from helping increase Johnnie Walker's sales by 48 percent in just eight years.[7] In other categories, product innovation is often the means by which brands seek to achieve advantage. This type of advantage is often short-lived, but, as Murray puts it, "the challenge is to sustain the feeling of difference." And sustaining that feeling is often easier than you would expect. Not only is it very difficult for competitors to displace a well-established idea, but people also equate "first" or "original" with "best."

3. Fighting for More than a Fair Share of "In-Play" Consumers

One of the biggest contributions that marketing makes to a business is simply holding the market share gained as a result of disrupting the status quo. After a disruption, markets tend to return to a state of dynamic tension, in which the actions of each brand tend to cancel out those of their competitors. Buyers lost as a result of dissatisfaction or the competitors' tactics must be balanced out—and, ideally, exceeded—by the acquisition of new buyers through the brand's own marketing and sales tactics. In the absence of real product or positioning innovation, top-line growth becomes a long, slow process of incremental gain. Even in fast-growing markets like

China, sales gains will lag behind those of the category if you don't hold market share.

At any one point in time, there are only so many people likely to buy within your product category. For instance, according to Polk, Americans are now keeping their new cars for an average of six years, two years more than before the Great Recession.[8] Most people driving a car today are not actively thinking about buying a new one. And when it comes to buying a new mobile phone, there is substantial variation by country. In the United States the replacement cycle is short, just under two years on average, but in many other countries the cycle is longer than three years.[9] Obviously, food and drink products are bought more frequently, but even so, not everyone is interested in buying all the time.

This is an important point that many people tend to overlook. For a large percentage of product categories, only a minority of people are "in play" at any one time. Those individuals who are actively considering a purchase or shopping within the category will be most likely to note, consciously reflect, and act on marketing communications and sales activities. Those not actively shopping within the category, either because the category itself is not relevant to them or because they do not plan to buy in it again for a while, are unlikely to reflect on what they see and hear. If the news is compelling enough, they might remember it when they do purchase within the category, but more likely, the specifics will simply be forgotten. In most cases, all that will be retained are a few general impressions. Marketing's job is to make these impressions as clear and compelling as possible so that when the time comes for these people to buy, they are positively primed to respond to the brand.

The Red Queen Effect

The term *Red Queen Effect* was first coined by Leigh Van Valen, an American evolutionary biologist.[10] The phrase refers to the Red Queen's pronouncement to Alice in Lewis Carroll's *Through the Looking Glass* that "it takes all the running you can do to keep in place!" Though the Red Queen was running, she was going nowhere, because everything around her was running too.

As applied to business, the Red Queen Effect describes situations in which companies need to escalate investment to maintain their market position while the returns on those investments are diminishing. Many suggest that Red Queen competition results from management

➡

preoccupation with operational effectiveness and cost control at the expense of strategic differentiation. Rather than looking for new ways of competing, companies fall back on doing what they have always done, just faster and more cheaply. Competitors respond in kind, thus creating a vicious cycle of quicker product cycles and falling prices. This cycle often continues until a third party disrupts the category by introducing a more innovative approach.

I believe that exactly the same effect applies to brands within a product category. As soon as one brand introduces a new and disruptive innovation to a category—whether tangible or intangible—and appears to benefit as a result, then its competitors will inevitably follow in its footsteps, thus negating the competitive advantage of the first mover.

We can see the Red Queen Effect in action in one of today's most rapidly evolving product categories: mobile phone handsets.

In 2005, the two big players in the U.S. consumer phone market were Nokia and Motorola. According to our estimates from BrandZ, these two brands held roughly half the category between them, with Samsung and LG trailing by a wide margin. BlackBerry, introduced in 2003, had yet to make significant inroads into the consumer market and was still the go-to business device. Market shares in 2006 remained mostly unchanged. BlackBerry and LG gained some ground. Nokia and Samsung lost some.

Then came Apple. In June 2007, the company launched the iPhone, and the fortunes of the existing brands changed dramatically. By 2010, the iPhone had consolidated its position as the most desirable (if not the most affordable) mobile phone. At the same time, BlackBerry had succeeded in making inroads into the consumer market. Both brands were more likely to be perceived as different from existing consumer models: they set the trends for the category and were more appealing than other makes. In terms of market share, Nokia and Motorola were particularly hard-hit, while Samsung and LG managed to hold their ground.

What was important about the introduction of the iPhone was that it reset expectations of what was important when buying a mobile phone. This was not just a matter of form and function. Which brand you owned had become more important in the decision. Before the launch of the iPhone, 35 percent of mobile phone buyers interviewed in the BrandZ survey said brand choice was very important. This number rose to 47 percent in the two years after the launch.

➡

Since 2010, however, the proportion of people saying brand choice is important has returned to 2006 levels, and the proportion saying brand is not important has risen back to 19 percent. Why? Because competitors have adjusted to the new standard set by the iPhone. The Samsung Galaxy S, first launched in June 2010, is now considered by many to be a direct competitor to the iPhone. Everyone, it seems, has their sights set on beating Apple at its own game, and, with many phones that appear functionally similar, which brand of phone you carry has once again become less important. The iPhone was once a game changer. But now the game has changed, and unless Apple can invest more and move even faster, it is bound to find itself responding to the increased competition.

Apple has tried to preempt competition by quickly introducing new models and dropping the price of the iPhone over time. Early in 2013 in the United States, a 16-gigabyte iPhone 5 cost $199, compared to $599 for the 8-gigabyte first-generation model. It has become Apple's modus operandi to start out by pricing new products at a premium (in part to recoup the investment in new manufacturing techniques) and then price down over time. If the company did not do this purposefully, then it would be forced to do so anyway. Outside of the United States, an iPhone 5 can still cost the equivalent of $650, and even the cheaper models are more expensive than Android phones. This high price did not stop Apple from reporting record sales of iPhones in the last quarter of 2012. However, the 28 percent increase over the prior year's sales failed to meet analysts' expectations, and, with profit flat, the company's stock price slid lower. Apple may not need to launch a cheaper phone to meet consumer demand, but it seems likely that it will do so to meet investor demand.

CONNECTING THE MEANS WITH THE END

In many companies, a fundamental disconnect exists between senior management and marketers regarding marketing priorities. One Economist Intelligence Unit report finds that "non-marketing executives—including the CEO, CFO, CIO, other functional heads and board members—see marketing's top priority as driving revenue, by a wide margin over finding new customers (30% to 19%). For CMOs, however, marketing's priorities are creating new products/services and customer acquisition; driving revenue ranks third."[11]

The two groups appear to possess very different perspectives on what is important, but I suspect that management is more focused on the end result—driving revenue—and marketers are more focused on the means—how to drive that revenue. Of course, both groups ought to be more concerned with driving profit growth than revenue. The value of strong brands lies not only in their ability to create demand but also in their ability to justify a price premium.

REDEFINING THE ROLE OF MARKETING

The biggest challenge that brands face is creating sustainable and profitable growth, but today, the vast majority of brands trade on past equity and transient buzz. As a result, consumers increasingly perceive brands as substitutable, and they value getting a good deal more than a specific brand. A widely held excuse for this situation is that it is impossible to sustain differentiation in today's competitive marketplace. As a result, many marketers tend to focus on creating awareness without regard to meaningful differentiation. They value innovation for innovation's sake; fame, not substance; buzz, not advocacy.

To my mind, the belief that it is impossible to sustain differentiation is an excuse. While product innovation remains the best way to create meaningful differentiation, it is certainly not the only way. Johnnie Walker is just one example of a brand that differentiated itself through effective marketing communication of a meaningful idea. There are always opportunities to change the way people feel about your brand, to change their expectations of the product experience, to introduce new criteria that separate it from the competition, to demonstrate how much others enjoy it, or to reframe price perceptions to make the brand seem like a good value.

MARKETING, REINVENTED

A survey of 191 organizations conducted by Forrester Research and Heidrick & Struggles finds that the majority of marketers (86 percent of those interviewed) believe they have ownership of brand strategy and positioning, creative development, and advertising.[12] However, less than a third felt they had ownership of pricing (31 percent), product and service development (30 percent), or in-store or branch experience (25 percent). While a higher proportion felt they had involvement with these functions, only 14 percent agreed that everyone in the organization had a synchronized view of the customer. How, then, can marketing be held accountable for revenue growth when some of the most important tools are not under their control and, worse, there is a lack of alignment regarding what customers want?

As I noted before, customers make no distinction between the brand they see in-store, the brand with which they chat online, and the brand they see in TV commercials. If a brand is to drive as much value as possible, then the first step must be to align all functions to deliver a consistent experience across all touch points. If marketing is to be held accountable for driving top-line growth, then it must move beyond its traditional boundaries and encompass all aspects of the customer experience, including product and service formulation, customer service, and point-of-sale activity. This requires not only a shift in the way marketers are perceived, from spin doctors to experience architects, but also a shift toward marketers taking responsibility for a much wider remit. They need to seek opportunities to shape the whole brand experience to be meaningfully different from that offered by the competition—instead of just worrying about their latest campaign.

Chapter 2

How Brands Influence
Purchase Decisions

Many people—particularly those outside marketing—often overestimate
the importance of brands to their consumers. Brands do serve as important
signals of quality, meaning, and value, but on a day-to-day basis, people
rarely invest much time and thought in them. Brands become most important when people need to make a decision between alternatives, and this is
when marketing can have one of its strongest effects.

Our understanding of how humans think and make decisions has
evolved considerably over the past decade. What is becoming clear is that
our decision-making is far less lucid and well-reasoned than we might like
to believe. Since the ultimate goal of marketers is to influence purchase decisions on a consistent basis, it is important to ask ourselves whether or not
our new understanding actually changes what we need to do to achieve this
goal. I believe that in most cases, the new learning simply confirms established marketing principles that unfortunately are often ignored. Far from
changing the practice of marketing, this new learning makes it all the more
important that we execute proven strategies and tactics as effectively as possible. It sharpens our focus on what really matters.

ONE THOUGHT AT A TIME

There is now compelling evidence to suggest that our ultimate decisions
about anything are based on little conscious thought. How little thought?
The available evidence used to suggest that we could simultaneously consider seven items, plus or minus two, in making a decision. In recent years,
however, the evidence has begun to suggest that we can really only consider
as many as four items at a time, and perhaps as few as one.

New research from the cognitive neuroscience laboratory at the French
National Institute for Health and Medical Research (INSERM) in Paris finds

that we can keep two goals in mind at one time by dividing them between the hemispheres of the brain's medial frontal cortex. The research, however, does not imply that we can actually work on both tasks at once. Etienne Koechlin, co-author of the study, writes in an email to *Scientific American* that "this finding further suggests that the frontal function cannot keep track of more than two goals/tasks at the same time. Humans have problems deciding between more than two alternatives....A possible explanation is that they cannot keep in mind and switch back and forth between three or more alternatives."[1]

Other research also suggests we may only be able to consciously focus on one idea at a time. Rather than being able to hold several ideas in our mental workspace, we may only be able to consider them sequentially. A study by Brian McElree, a psychology professor at New York University, supports this hypothesis. When presented with a string of numbers, people could recall the immediately preceding numbers in a fraction of the time it took them to recall numbers presented more than one number earlier.[2] McElree states, "There is clear and compelling evidence of one unit being maintained in focal attention and no direct evidence for more than one item of information extended over time."[3]

If that finding is true, then it has profound implications for how we make conscious decisions, since any decision then becomes the outcome of the idea held in our minds at the time we made the decision. All the more reason for marketers to try to make sure that the most relevant, motivating, and differentiating idea of a brand comes readily to people's minds when they are faced with a decision about purchasing that brand. But not every decision we make is the result of conscious deliberation. In the interest of physical survival, we have evolved to ensure that our brains are working only on the things that really matter. We rely on a second way of making decisions that parallels the one used by the deliberative system and makes decisions quickly and easily based on our prior experience.

QUICK DECISIONS

In his book *Thinking, Fast and Slow*, Daniel Kahneman defines our two ways of thinking as System 1 and System 2. System 1 is fast, automatic, and instinctive. System 2, which occurs in our conscious mind, is slow and deliberative.[4]

According to Kahneman, the instinctive System 1 way of making decisions is informed by our previous experience and learning, and an important component of System 1 decision-making is the emotions that are evoked by a particular object or event. Emotions are far from irrational. Rather, they are important signals established by our prior experience that help us decide whether something is good or bad, something we should approach or avoid.

The decision-making process of the instinctive system is fast, far faster than the decision-making process that occurs in the conscious mind. How fast? How about one-twentieth of a second? That's how long it takes people to make a decision about the visual appeal of a website.[5] There is, of course, a vast difference between product categories in terms of the typical amount of time and deliberation that people devote to making a decision. When craving a snack, we quickly decide between salty and sweet and instinctively reach for a brand with the desired taste. How salient and visible a brand is at the point of purchase and our instinctive response to it will have a big influence on whether or not we choose it. Items in the grocery store are more likely to be bought based on habit, and most of the time the decision is made with little conscious deliberation.

Gerd Gigerenzer of the Max Planck Institute for Human Development in Berlin proposes that all human decision-making is done using what he calls "fast and frugal heuristics" to reach conclusions that are satisfactory and comparable to conclusions reached through more deliberation.[6] These heuristics are fast in the sense that they facilitate quick decision-making. They are frugal in the sense that they allow decisions to be made based on limited information. Critical to this discussion is the idea that fast and frugal heuristics help us make choices in spite of the fact that we lack perfect knowledge and infinite time. Simple heuristics or rules of thumb guide our decision-making, with the simplest one being, "I'll buy the one I bought last time."

Forget the myth that 70 percent of decisions are made in-store. All decisions are made in-store; it's just that the vast majority are made with little conscious deliberation. Instead of thinking about which brand of toothpaste, cereal, soft drink, or detergent to buy, we choose the brand we bought the last time (provided our needs, brand knowledge, or satisfaction with the brand have not changed). When it comes to this type of habitual purchasing, cues are critical. Whether they have to do with the position of brands on the shelf, the color of a box, or the shape of the bottle, cues are the triggers that set our "automated" purchasing into effect. Research by OgilvyAction finds that only one in ten shoppers changes his or her mind about what brand to buy in-store.[7]

Decisions made to buy brands in higher-risk, higher-price product categories—like cars, health care, or financial services—are more deliberative than decisions made to buy packaged goods. Most people try to consider the merits of the different options before making a choice. But their conscious deliberation is not independent of their instinctive reaction to the different choices (assuming that they even have all the facts they need to make the decision and time to consider them). A positive instinctive response to one brand will predispose people to seek out reasons why that brand is a better choice than others. So while the instinctive response may not triumph over conscious consideration—if nothing else, the desired brand may be unaffordable—it still has an important influence on our deliberations.

What evidence do we have that our emotional response to brands shapes our purchasing behavior? Since 1994, Millward Brown has consistently found that advertising that is emotionally engaging and linked to the advertised brand in people's minds produces both short- and long-term sales effects. More recently, analysis of 880 IPA Ad Effectiveness Award entries by Les Binet and Peter Fields found that advertising that is more emotionally engaging outperforms rationally based campaigns by a factor of two in terms of profit effects over a three-year time frame.[8] This effect is not apparent in the short term, because people shopping a product category today are more consciously attuned to think about the relative merits of the different brands. But people who are not in the shopping mode are more likely to be swayed by emotionally engaging advertising even if it does not make them think about buying the brand then and there. This predisposition primes them to consider the brand when they do come to shop the category.

WHY SMALL DIFFERENCES CAN MATTER TO MARKETERS

In a presentation given at a client workshop I attended, Nick Chater, a professor of behavioral science at Warwick Business School, proposed that humans find it impossible to analyze anything in isolation.[9] Instead, we tend to compare things to very close alternatives. As a result, context and comparison are crucial to our ability to make decisions. Given this tendency, Nick concludes that you should make sure your brand is not comparable to any other. If it is comparable, then people will either give up trying to decide between similar alternatives or default to choosing on the basis of price.

Heuristics theory states that you need to provide people with a reason to choose your brand, just as Nick Chater suggests. Heuristics provide convenient shortcuts and, most important, clear stopping points for decision-making. Humans are simply not able to handle complex sets of variables and make totally rational decisions—whatever classical economists might try to tell us. In the absence of some obvious and important factor, the reason we use to make a choice need not be significant; all else being equal, it merely needs to be differentiating enough to tip the balance in favor of your brand. It could be a feature, a feeling, or an impression, but it needs to differentiate your brand in a positive way, and you need to ensure that potential customers are familiar with that difference.

Habit and our instinctive response can only have an overriding influence on decision-making when our experience gives us a relevant frame of reference. If we are making decisions about something familiar and easily recognizable, then it is unlikely that our conscious thought will extend much beyond "that's it." However, when circumstances change or we are faced with a completely new decision, our habits are broken and our feelings are less compelling. In these circumstances, our experience provides less direction, and the influence of emotion is reduced, so conscious thought comes into play. And if opportunities for conscious thought are scarce, then we need to make the most of the odd occasions when people do consciously think about their purchase decisions.

JUSTIFYING A PURCHASE DECISION

It is when people are forced to make choices between alternative brands that claims become important. As noted previously, although people may not be conscious of it, they are drawn to facts that justify their positive feelings about a brand. Peter Ayton, a professor of psychology at City University London, puts it this way: "We seem to value the appearance of having good reasons for our decisions more than we value the reasons themselves! Choice between dissimilar alternatives will be easier to justify with reasons than any choice between similar alternatives."[10]

In other words, people like to believe they are rational decision-makers even when the scientific evidence suggests that they are not. Graham Page, executive vice president of Millward Brown's Consumer Neuroscience Practice, states: "One of the most important roles of advertising is to give people an easy justification for their purchase decision. It's easy to forget that being rational feels good. Even if our decision-making is based largely on instinct, we don't like that idea and would prefer to have a ready explanation for our choices."[11]

REHEARSE AND REPEAT

If you are the incumbent brand, then the last thing you want is for someone to reconsider which brand he or she should buy. The objective should be to make your brand the default choice and stop people from thinking too much about their decision. Because if people do need to think about it, then there is a good chance they will either choose another brand or end up dissatisfied with their choice.

So how do you help someone have an instinctive, good feeling about your brand? One of the most important things that advertising can do to build a brand is to help people "rehearse" what the brand experience will be like—that is, to portray how a brand is used and what it feels like.

Take the example of Maltesers, a confectionary product made by Mars. Maltesers are spherical balls of malt honeycomb covered with milk chocolate. For many years, they were advertised in the United Kingdom as being less filling (and, by implication, less fattening) than ordinary chocolates. A new campaign launched in 1999 by DMB&B sought to give the candy a more lighthearted and fun positioning and, in one ad, powerfully demonstrated what the previous advertising had been saying for years. In an ad titled "Feed Me," a couple is lying together on the sofa. She demands him to "go on." He obliges by sucking a Malteser onto the end of a drinking straw and popping it into her mouth. Without saying a single word, the ad demonstrates just how light Maltesers are, while at the same time implying their irresistibility, through a rehearsal of the brand experience.

You may not think that you need much rehearsal to recognize a brand, but the latest scientific evidence suggests that we do not store coherent "images" of brands (or of anything else, for that matter) in our heads. Our perceptions of the world are not the detailed observations we think they are but are constructed on the fly from very limited information. So every time we encounter something related to a brand, various types of stored associations bubble up from different parts of our brains into our mental workspace and form an image of that brand. The more quickly, vividly, and completely this mental representation forms, the more likely we are to consider the brand for purchase.

Implicit in the use of the word *rehearsal* is the concept of repetition. Although there is no convenient rule of thumb for how many exposures an ad should get, we do know that repetition matters, and now we have a scientific rationale for why it matters. The evidence suggests that the more frequently we draw on our mental associations to form a representation of a brand, the stronger and more complete the representation becomes. This repeated mental rehearsal of what the brand is about helps build clarity and maximizes the chance that the brand will be recognized and found desirable.

In their award-winning paper "Cognitive Neuroscience, Marketing and Research: Separating Fact from Fiction," Professor Jane Raymond of the

University of Wales, Bangor, and Graham Page describe how brand representations are formed from three different types of information:

1. Cues that help us recognize the brand
2. Knowledge about what it is like to use or experience the brand
3. The emotional value the brand holds for us

Exposure to the brand—through use, advertising, or word of mouth—has the potential to strengthen, augment, or modify our existing associations. Marketers have always valued the first type of association (brand cues) and have come to appreciate the third (emotional value), but they often ignore the importance of the second (experience). Yet some of the most effective advertising is effective precisely because it creates expectations about the brand experience.

Advertising has a unique value to the marketer in that it can portray a positive view of a brand without including any of the negative associations that invariably accompany real-life experience. It can do this in a variety of ways: by ensuring that potential customers have a clear idea of what the brand looks like, by setting up expectations of what it will be like to use the brand, or by establishing a positive impression that makes the brand more desirable.

If the advertising portrayal is believable, then brand users are likely to focus on the aspects of the brand experience that justify their belief in the brand, and as a result, their predisposition toward the brand will be strengthened.

ENHANCING PERCEPTIONS OF BRAND EXPERIENCE

This process of framing peoples' expectations of the brand experience is one aspect of what Gordon Brown, one of Millward Brown's founders, terms *enhancement*. A keen observer of human motivations and how advertising works, Gordon recognized the power that rehearsal can have in creating brand perceptions. Speaking of product categories with little differentiation in a paper written back in 1991, he observes: "Vivid advertising memories powerfully enhance perceptions of the product field benefit, which in reality all brands would deliver! A woman who tries a brand of furniture polish with vivid advertising memories demonstrating that it gives a deep shine will see a deep shine."[12] Of course, as he goes on to say, all polishes are designed to shine, but the advertising manages to condition the buyer's encounter with the brand to focus her attention on the shine produced by that particular brand.

Gordon goes on to talk about categories in which different brands offer different benefits. He uses the example of laundry detergents, which may work in cold water, be fragrance free, or be effective at removing stains. In categories like this, Gordon says, "Advertising memories cause consumers to 'test out' the attributes that are in reality the product's strengths, diverting them from the weaknesses!"[13]

Gordon also confirms the importance of creating memorable and enduring brand associations, stating that "consumers make up their minds when they think about brands. This is mainly when they experience brands."[14] In other words, advertising can cause people to anticipate (or remember) what it is like to use a brand, and the reality of the experience—shaped by the advertising depiction—will determine their lasting impression of it. In a unique experiment conducted in 1994, Andy Farr of Millward Brown demonstrated that exposure to both a brand's advertising and its product had a far greater effect on future purchase intent than either seeing the advertising on its own or trying the product without seeing its advertising. Farr concludes:

> [The experiment] powerfully supports the view that traditional theme advertising—the repetition of relatively weak arguments, creatively presented—does not influence product perceptions at the time it is seen, but increases the likelihood of exploratory purchases—out of "interest"—later, and the advertising memories have a powerful influence on whether the product is felt to be good or bad competitively, when it is tried, increasing the chances that the advertised brand will be purchased subsequently.[15]

Brand experience exerts the single biggest influence on brand associations. By confirming or denying expectations, experience determines our beliefs about a brand. However, those beliefs are often influenced by impressions gathered prior to purchase through word of mouth, search results, and advertising. Even nonusers experience brands by seeing them on the street, hearing people talk about them, and so on. In some cases, experience will supersede expectations, but in most cases, our expectations shape how we experience a brand.

At best, a brand impression conveyed by advertising and *not* supported by experience such as use or seeing it in-store or on the street is likely to have only a transient effect.

SALIENCE MATTERS

I mentioned that the more quickly a mental representation of a brand forms and the more vivid and complete that representation is, the more likely it is that a person will consider the brand for purchase. This speed with which a representation comes to mind is known as *salience*. Salience has an impact in two ways.

First, it is a major factor in determining how readily a brand comes to mind when people think about a category. In every category, there is a strong, positive relationship between salience and purchase. Just being the brand that comes to mind first can be a tremendous advantage, particularly for brands that are bought on impulse, and that advantage persists throughout the purchase process.

Second, salience helps explain why people are behaviorally loyal to a brand. People are more likely to bond with brands that are salient in terms of attributes that are important to them. So if "helps prevent decay" is an important attribute for toothpaste and someone associates that attribute only with Colgate, then he or she will be more likely to buy Colgate than alternative brands.

If your brand is the most salient, it can be really tough for the competition to undermine that positioning. No matter what they say or do, people will tend to consider your brand for purchase first (see the boxed section below on Double Jeopardy). But that does not mean that leaders can maintain that advantage just by being big. MySpace was pretty big, but Facebook is now far bigger and MySpace is struggling. BlackBerry was not the first personal digital assistant (PDA) with email. Anybody remember Palm? Time after time, the evidence demonstrates that being the first brand to create and lead a category does not mean you will keep that leadership. Google was not the first search engine, but it was demonstrably better than the ones that preceded it.

Double Jeopardy and Its Implications

Double jeopardy is an empirical generalization of a phenomenon that has been observed across a multitude of product and service categories: the greater a brand's penetration, the higher its repeat purchase rate. This means that most of any brand's market share can be explained by its penetration and relatively little by its repeat rate. Double jeopardy appears to have the most sway in categories in which brands are readily substitutable and differentiation is low.

The double jeopardy relationship also applies to attitudinal data collected in surveys. For instance, in Millward Brown's tracking and brand equity data, better-known brands show more than their fair

➡

share of recent usage and consideration. In order to predict actual market shares, stated purchase intentions must be adjusted to take account of the double jeopardy relationship. People who claim they will buy a small brand are less likely to fulfill their intentions than those who claim they will buy a big brand. When looking at brand image data, it is apparent that better-known brands attract higher levels of endorsement on positive attributes.

Double jeopardy is a fact of life. Big brands have a significant advantage over small ones. The key question is: What do you do with that information? One important conclusion, according to Professor Byron Sharp and his colleagues at the Ehrenburg Bass Institute, is that marketers need *reach*; that is, they must talk to light buyers and nonbuyers if they want to grow market share.[16] Actions that increase a brand's mental and physical availability will drive market share growth. These findings are incontrovertible, although many marketers seem to ignore them. Unfortunately, in his book *How Brands Grow*, Sharp also proposes that this link between availability and growth means that "distinctiveness" is more important for a brand than "differentiation." He states, "Rather than striving for meaningful, perceived differentiation, marketers should seek meaningless distinctiveness. Branding lasts, differentiation doesn't."[17]

This "advertising as publicity" model may seem bizarre to people brought up on a diet of "advertising as persuasion," but in fact there is a great deal of truth to this argument. For established brands, particularly those in low-involvement and habitual purchase categories, it is true that much of what advertising does is to reinforce existing brand memories and impressions. Creating a positive halo around the brand does influence future purchases. Research finds that over the long term, brands that have effective advertising exposure greater than their market shares tend to grow.

However, the argument breaks down when we shift our focus to new brands and brands that actually have a new and compelling message to convey. When people encounter information that is new and personally relevant, their perceptions of brands (and, by implication, their perceptions of competitive brands) can be changed. Therefore, advertising that conveys new information can have a substantial effect on sales. However, in most categories, the advertising that generates the strongest response combines all these attributes: it is well branded

➡

and highly memorable, *and* it evokes a conscious, positive response to the impression delivered by the ad.

Now, of course, such a combination does not occur that often. As Sharp notes, "new news" is only newsworthy for a limited period of time. Thereafter, upon repeat viewing, an ad will remind people of why their opinion changed, but it won't create further change. The challenge for marketers, then, is to continue to present new innovations or to make old news seem new again in order to refresh what the brand stands for in people's minds. In the absence of new and compelling news, advertising becomes a battle for salience, and ensuring that creative memories are linked to the brand in people's minds will always prove the winning strategy.

If you're the leader, the trick is to be alert and attentive to what your customers want—and then make sure you're giving it to them. If you're a challenger brand, your job is to understand the status quo and then create a new relationship. You may win consumers over by offering them a different reason to choose your brand or simply by proving that you're better than the leading brand. You have to win consumers' hearts by showing off your brand's best characteristics and making sure they don't simply forget or ignore them. Give consumers a reason to look twice.

WHAT DOES YOUR BRAND REALLY STAND FOR?

Many pundits suggest that brands need to form bonds with their consumers, encouraging people to love the brand just as if it were a person, but I would suggest the bond between brand and consumer is strengthened as much by inculcation as infatuation: a brand becomes firmly fixed in a consumer's mind as a result of repeated experience. In other words, bonds form when a brand becomes salient.

Saliency is context-specific, and this is where brand associations come into play. Take the following example. I am driving north on Interstate 95 to Millward Brown's Fairfield office. Where should I stop for a quick bite of breakfast? Invariably, I end up at the McDonald's drive-thru in the service area at Exit 22. Why don't I get off at Exit 21 and go to Starbucks? Because Starbucks does not come readily to my mind when I think of breakfast.

Of course, when I am driving up the interstate to the office, no such "discussion" ever takes place in my head. Making conscious, thought-out decisions takes time and effort, and everyday decisions such as where to get breakfast don't warrant that much energy. So, like everyone else, I shortcut the decision process. When the key words summing up my need are "breakfast food fast," McDonald's instantly becomes the destination. Do I love McDonald's? Not really, but I have eaten at that McDonald's on I-95 more times than I would care to admit. (I do remember, though, when the first McDonald's opened in Leamington Spa, England. We thought it was very cool at the time!)

The famous ad man Maurice Saatchi made a keen observation about the power of associations in an article in the United Kingdom's *Financial Times*. Maurice would have us boil a brand idea down to one word: "In this new business model, companies seek to build one-word equity—to define the one characteristic they most want instantly associated with their brand around the world, and then own it. That is one-word equity."[18]

What Saatchi describes is the modern equivalent of the best location on Main Street, except that the location is in the mind. I am not sure that the association needs to be one word, but it does need to be one idea, and that idea should be crystal clear. Importantly, however, the idea does not need to appeal to everyone. In fact, a more focused idea is more likely to appeal to the brand's real target. And that, of course, is where so many brands go wrong. Too often brands try to extend their appeal to something that does not fit their core idea. In trying to extend their appeal to a new audience, category, need, or time of day, they undermine what the brand stands for.

For Starbucks, breakfast was an afterthought, seemingly just another way to drive revenue. For me, that extension diluted the core brand idea. What's breakfast got to do with a comfortable place to relax over a decent cup of coffee? What was Red Bull thinking when they launched a cola? To most people, Red Bull equals energy, period. The brand should quit while they are ahead and be content to dominate the energy drink category. However, Apple equals different. That idea allows you to go a lot of different places—computers, music, phones, and so on.

IMPLICATIONS FOR MARKETING

Thinking about all of these ideas from a marketing perspective, there are several conclusions that I would draw.

First, you want to encourage people to develop a positive instinctive feeling toward your brand. The more positively they feel about the brand and the more salient it is, the more likely they are to choose your brand without intense deliberation. The best way to encourage positive feelings about a brand is to get people to experience the brand. Sampling, test-drives, and demonstrations are all tactics that encourage people to physically experience a brand, but they are all limited in reach. Advertising has the ability to create powerful and compelling portrayals of what it feels like to use a brand and create an emotional halo around it at the same time.

Second, you want to make sure that people have ready justification for their instinctive feelings in case they are forced to reconsider their brand allegiance. Professor Ayton states, "People look for preferences that are easy to justify."[19] So it helps if people can have a nice, simple, and easily remembered claim to support their decision-making. Facts that help justify people's decision to buy the brand are important, not because they motivate purchase, but because they justify it. But there is a caveat: those facts need to be consistent with the instinctive impression people have of your brand, which requires consistency in messaging.

Third, if you are seeking to change people's minds at the point of purchase and do manage to counter someone's instinctive desire to buy another brand, then you may risk buyer dissatisfaction with the choice over time unless people's experience of the product rapidly exceeds expectations—that is, unless you can create a positive instinctive reaction to your brand. Focusing attention on the most compelling and distinctive aspects of your brand will help people focus on its strengths and ignore its weaknesses.

Fourth, many marketers seem to have forgotten that people typically buy brands to meet a specific need. That need might be to solve a functional problem, restore a positive mood, or improve their social standing, but in each case, the acid test of whether someone will buy the brand again is how well the experience lives up to their expectations. Does the brand deliver on its promises? Influencing how people anticipate and experience the brand is a powerful means to sway purchase decisions, but all too often, marketing campaigns end up chasing superficial trends and not focusing on the real reason that people buy brands.

The most effective marketing is rooted in the brand experience—it's not tangential to that experience. We need to reframe or enhance the feeling of what it is like to use the brand, not simply draw attention to the brand. Rehearsal is a critical mechanism for navigating life and marketing alike. Rehearsal combines imitation and repetition, both of which are critical to learning new skills or improving on existing ones. While advertising cannot replicate the exact events through which someone will experience, or has experienced, a brand, it can represent that experience in such a way that it is easily recognizable and relatable, allowing people to imagine themselves as the protagonist in the event.

Chapter 3

What Makes Your Brand Meaningfully Different?

For many years, a smooth green stone has sat on my desk. It's a piece of serpentine that I was given when I visited an artist's workshop in Scotland as a small child. Truthfully, it's a pretty unremarkable rock, and I doubt that anyone else would find it interesting, but it means something to me. People have always been drawn to ascribe meaning to inanimate objects—often more meaning than is justified by those objects' functional benefits.

Look up the word *meaning*. Its synonyms are *sense*, *gist*, *connotation*, *import*, and *significance*. In the context of a brand, meaning has nothing to do with features and functional benefits and everything to do with people's understanding of what the brand stands for, the sense that the brand has something unique to offer. Strong brands evoke an instinctive attraction and offer a ready justification for that attraction. Rather than try to figure out which brand is best, people are drawn to the one they find most meaningful.

Without meaning, brands would have no value; everything would be a commodity, and a physical or tangible advantage would eventually erode as competitors copied, blunted, or superseded it. Meaning is far more subtle and elusive. If it is difficult to pin down and identify, then so too is it difficult to copy.

Brand meaning can originate from a multitude of sources. It could come from your personal history with a brand: you might use the same brand of detergent that your mother did. Or it could come from functional characteristics: you might really like the intuitive interface of that tablet computer. You might be attached to your car because you think it looks hot, or because it is economical and saves you money. Or a brand's meaning for you might simply be that it is familiar whereas others are not. Meaning can be functional and tangible or emotional and intangible or all of the above. Meaning is in the eye of the beholder.

◼◼◼ THE VALUE OF MEANINGFUL DIFFERENCE

It is not enough for a brand to be meaningful. Successful brands are both meaningful *and* different. As we noted in the last chapter, we humans find it impossible to judge anything in isolation; instead, we tend to compare things to very close alternatives. So a *difference*, a factor that distinguishes one item from another, gets our attention. And while a difference may be apparent to most people, it won't seem important to everyone. A *meaningful* difference is one that is considered to be important—one that provides a brand with a meaning that is likely to have an influence on a person's brand choice.

Our research finds that people are more likely to buy a brand when they find it to be different in a "good" and "meaningful" way. The same characteristics help a brand command a price premium and create the potential for the brand to grow future market share. We will examine these relationships in more detail in Chapter 4.

◼◼◼ WHAT IS A MEANINGFUL DIFFERENCE?

A meaningful difference is one that is significant and influential. It might be a tangible product-oriented difference, or it might be an emotional and intangible difference. A difference that may seem trivial to some people may for others add that extra something that makes them choose that brand over the alternatives. A brand like Coca-Cola is unlikely to change the formula for its syrup on a regular basis. Instead, Coke works single-mindedly to create intangible differentiation through creative campaigns that focus people's attention on the brand's reason for being. The brand's "Open Happiness" advertising, "My Coke" Rewards, and engagement with social media all remind people of what Coke stands for. The company layers new and renewed meaning onto a long-established and dominant brand. As a result of this communication, ardent fans continue to believe that there is simply no substitute for their favorite soft drink.

Small differences, even intangible ones, can have big effects in relatively undifferentiated categories. The Old Spice campaign "The Man Your Man Could Smell Like" was different not only from previous Old Spice campaigns but also from any other existing campaigns for male grooming products. In pushing the brand into new territory, the campaign risked being seen as inappropriate to the brand and category, but instead, its tongue-in-cheek humor managed to engage a new young audience, thus reframing the brand and boosting its growth. The Old Spice campaign didn't work by making people note the brand's "relevance" and immediately add the product to their shopping lists. It simply gave them a fleeting moment of amusement. But later, when they noticed the brand on the shelf, that amusement, whether consciously remembered or as part of their instinctive response, made the brand different—and for some people, that difference was enough to make them choose it.

An intangible difference rooted in advertising memories can be enough to get someone to try a brand. But to finish the job of winning over a new user, a brand has to deliver a rewarding experience. A positive experience will confirm the user's belief in the brand's unique value, perhaps to the extent of supporting a higher-than-average price.

In this way, I believe that brands are like my lump of serpentine. They start out with a meaning that is fresh, new, and specific, based on when and how we first encounter them. Over time, familiarity and experience add new layers of significance. Some of that added significance may be quite tangential to what originally motivated our interest in the brand, but it's likely to be important to us nevertheless, just as the meaning that has accrued to my lump of serpentine over the years is important to me. So one of the most critical things a marketer can do is to continue to add meaning to a brand over time through events, sponsorships, and compelling communication. It is when a meaningful difference is conveyed and delivered in a way that resonates with consumers that attitudes and behaviors can be affected.

MEANING, NOT RELEVANCE, IS THE DIFFERENCE MAKER

When we study today's most profitable and successful brands, the importance of meaningful difference becomes obvious. The world's strongest brands are not the ones that are most relevant. On the contrary, the most successful brands don't try to be all things to all people. Apple's shares of the computer and mobile phone markets are still relatively small, held in check by the price premium Apple commands. But Apple sits at the top of the BrandZ Top 100 Most Valuable Brands Ranking. Burberry and Tiffany have both de-listed cheaper items that were attracting the "wrong" type of customer. And customers make their own choices as well; not everyone wants to shop at Walmart.

Many marketers today value relevance to the exclusion of difference—and to the detriment of their brands. Relevance alone is not a difference maker. It's necessary but not sufficient for brand success, and on its own, it won't motivate purchase, especially when brand switching is involved. Even when acting on impulse, people need to find something different and appealing in the brands they choose.

In some categories, like salty snacks, carbonated soft drinks, and confectionary, it may be enough for a brand to have built up a bank of positive memories in people's minds—that is, to be salient. A pleasant and rewarding association may be enough to prompt someone to say, "I like that one more than the others." In other categories, like computers, insurance, and airlines, marketing communications will need to provide a more coherent story of why a brand is different from and better than the alternatives, if only to reassure people that they are making a good choice.

In the absence of a meaningful difference between brands, the cheapest brand may be regarded as the best choice, because lack of differentiation turns brands into commodities and marketing messages into white noise. A meaningful difference can spark consumer interest and fuel demand for a brand, even when that brand carries a significant price premium. In today's complex, confusing, and increasingly impersonal world, people cherish meaning wherever they can find it, whether it's in a brand, a memory, or a lump of rock. So to build brand value, give people a reason to cherish your brand.

THE THREE LAYERS OF BRAND MEANING

The challenge for any brand is to identify a compelling meaning that will satisfy the functional and emotional needs of as many people as possible at a good margin. The understanding of a brand must be shared. In *The Global Brand*, I put it this way: "Any symbolic power wielded by brands is rooted in a collective understanding of what they represent. Perceptions of powerful brands such as Coca-Cola, Apple's iPod, and Harley Davidson consist of well-known and widely shared associations, which form a base on which people add their own individual, subjective reactions."[1]

To illustrate this idea, consider McDonald's. You only have to see the logo on a crowded street to know what the brand stands for. Whatever your personal thoughts about McDonald's, we all share a clear understanding that it delivers fast food at a reasonable price. Individual interpretations may differ in terms of exactly which aspect of that understanding comes to mind first. For one person, it may be "convenient, fast, cheap eats"; for another, "good for kids"; and for a third, "milkshakes," but each would understand why the others had interpreted the brand that way.

There are three very distinct layers of meaning to a brand—that is, how a brand is perceived—and marketers need to work differently to motivate people within each level.

To illustrate the three levels, let's think about a brand as an onion. This is a well-used analogy but pertinent nonetheless. I will start from the outside of the onion and work inward. Each layer encompasses the ones inside it and can influence or be influenced by the others. The first two layers deal with meaning that is shared across individuals. This cultural meaning is the broad, generational understanding and social meaning that is derived from a specific group or tribe within that culture.

Cultural Meaning

Cultural meaning is the broad context in which consumers come to appreciate what a brand stands for and respond to it.

The interaction between a brand and a culture is most obvious when the brand taps into the local culture directly. For instance, Molson beer in Canada, Walker's Crisps in the United Kingdom, and Guarana Antarctica soft drink in Brazil have all leveraged national pride to good advantage. Some brands are so strong that they are not just influenced by popular culture but also play a part in influencing that culture. Brands like Google, Facebook, Starbucks, Nike, Patagonia, and Innocent have shaped the world in which we now live and have earned unique status in the minds of many as a result. Success for these brands is not just a matter of jumping on the latest cultural bandwagon; rather, these brands shape and inform the cultural debate. They are built from a sense of purpose that permeates the entire brand experience. These brands resonate with their target audience because they are meaningfully different.

Social Meaning

The layer of social meaning is more specific and personal than the broad cultural context in which we live. In my mind, *social* refers to any group of associated people—colleagues, people who support the same club or pursue the same hobbies and interests, neighbors. Other people might refer to these groups as *tribes*, but that label implies a strong degree of allegiance, and I am not sure that you need to feel strong allegiance with a group of people to be influenced by them.

Professor Alex "Sandy" Pentland of the Human Dynamics Laboratory at the Massachusetts Institute of Technology (MIT) supports the "monkey see, monkey do" school of social science.[2] In an interview on Boston public radio, he recounts an experiment that tracked the voting intentions of people in a college dorm. One-third of people changed their political views based on face-to-face encounters. Overall, the people with whom the students spent time correlated with their voting behavior. In another case, Sandy suggests that if you hang out with people who routinely eat three slices of pizza, guess what? You, too, will start to eat three slices of pizza.

It should come as no surprise that we humans are incredibly susceptible to the behavior of others. In an effort to fit in, we sometimes consciously, and often unconsciously, model our behavior on the behavior of those around us. We want people to like us (and I do not mean just on Facebook) and accept us. In Western cultures that tend to be more individualistic, people are more likely to choose brands for what they say about themselves as an individual: choosing the brand becomes a badge of self-identity. In more collectivist cultures, the desire to fit in and be part of the crowd is still strong, and brands become a sign of affiliation and belonging. That sort of motivation is not dead in the West; it is just triggered by circumstance. If you are North American or European, picture yourself walking into a bar in a strange town where you know no one. Everyone in the bar is drinking the local beer. Might it seem like a good idea to order that beer as well?

The significance of social culture is most apparent when considering "identity" brands, or those brands that openly indicate something about a person's lifestyle and attitudes. Wearing a Patagonia t-shirt makes a different statement than wearing Billabong or Nike. In both developed and developing countries, a host signals both his own status and his respect for his guests through his choice of food and drink brands. Simply seeing the same brand being used by your close acquaintances can influence your own choices. I have been fascinated to watch the iPhone meme pass through a network of business associates until most of the people involved in the network were sporting the phone.

Individual Meaning

The third level of meaning is in the eye of the beholder. Although marketers add meaning to their brands, each person also adds his or her own unique meaning to what the brand stands for, and this individual meaning is far more motivating than the collective meaning. This is the domain in which small differences can have big consequences. A brand simply needs to be different enough for someone to choose it and pay a premium for it. The source of that meaningful difference may be so unique or so trivial that it is not readily appreciated—even by the person buying the brand.

We typically think about marketing as the creation of moments: communications, events, spectacles, launches, and so on. However, these moments make up only a very small part of the view customers have of a brand. The vast majority of what informs a person's view of a brand is the day-in, day-out usage of the product or service. That experience may not be thought about consciously that much, but it is the bedrock on which any brand is built.

In the documentary *Objectified*, Bill Moggridge, co-founder of the design firm IDEO, talks about the way that our appreciation of objects develops over time: "I like the concept of wearing in rather than wearing out. You'd like to create something where the emotional relationship is more satisfying over time. And...you may not worry about it or think about it that clearly, and people...don't have to have a strong love relationship with their things, but they should grow sort of a little more fond of them, perhaps, over time."[3]

I believe the same is true of brands. Fleeting and little-considered experiences accrete to produce an instinctive positive feeling toward the brand. Marketers should do all they can to reinforce the positive feelings created by those experiences. At the very least, how the brand is presented in marketing communications should be consonant with users' understanding and experience of the brand. However, for brands that have a one-to-one relationship with their customers, the opportunity exists to create new positive experiences.

In *The Global Brand*, I relate how Arc'Teryx replaced for free a ski jacket that I had gashed on the branch of a pine tree.[4] This positive experience has led me to buy more of their gear and extol the company's virtues on numerous occasions. But thinking back on it, the thing that really struck a chord with me was the phone call I received telling me that they would replace, not repair, the jacket. The person on the other end of the phone came across as positive and upbeat and clearly was not working from a script. While announcing, "We've never seen a tear this bad," she also managed to convey a sense of appreciation for the idiot who could do that much damage.

The most powerful connection is made when people believe the brand has their best interests in mind and is actively seeking to reward them for their loyalty. Equally, marketers need to do everything they can to avoid creating a negative experience for customers, because a poor experience will undermine people's feelings for a brand faster than anything else. For example, a bad experience might produce the following thought:

I used to think XYZ was a quality airline until the baggage handler broke my guitar.

Or,

I used to believe that XYZ bank was giving me a good deal until my friend told me about the interest rate that she was getting.

Or,

I used to love XYZ brand detergent because it seemed to have values I share, but then it let me down and ruined my best shirt.

The opportunity to interact with customers or consumers directly ought to be a good thing, and many companies aspire to build a one-to-one relationship with their consumers, but the BrandZ data suggest that it is actually more difficult for service brands to achieve high customer satisfaction scores than product brands. Direct interaction may be less something to be desired than a curse that exposes every weakness of your business.

So why worry about these layers of meaning? I think there is a real need for marketers to distinguish between cultural and social meaning—collective meaning that is relatively stable—and individual meaning, which, while more powerful, is far less stable among different people and over time. Marketers need to influence the cultural and social meaning to provide a stable context for people's personal brand experience. While inherently variable, that experience will be interpreted in the context of the general understanding of the brand. A classic example of this effect is described in an IPA case study of BMW written by Tim Broadbent. He quotes qualitative research describing how the brand helps BMW owners "forgive" any problems with their cars, saying, "If I'd had these experiences with a Rover, I'd think 'Typical.' As it's a BMW, I think I'm unlucky."[5]

Of course, failure to address the problem would override that belief, and so marketers need to make sure that the whole brand experience is as positive as possible. But it is equally important to create a positive groundswell based on a compelling and meaningful difference. That groundswell can create a shared perception that reinforces individual beliefs and helps the brand become part of the contemporary culture.

BEYOND FUNCTION AND FEATURES

Knowing where to focus your marketing efforts is in large part dictated by whether or not you believe you can create tangible differentiation, but I would strongly suggest that all differentiation needs to be rooted in the functional benefit of the brand if it is to be successful.

Most brands in established categories hit a wall when it comes to effective product differentiation. Take the example of digital cameras. For quite some time, there has been a megapixel war going on. While it is true that the number of pixels is related to the quality of the image, we have now passed the point at which most people can notice quality differences. Consider your own digital camera. Do you know (or care) if it takes the very best pictures possible? Or do you just want a device that is easy to use and captures your vacation memories? Lacking a meaningful and detectable product difference, many brands now seek to appeal to other differentiators: self-esteem, cultural empathy, and personalization.

Some brands, like Dove, Innocent, and Red Bull, have successfully engaged people on an emotional basis, and in Chapter 7, we will consider a range of ways in which brands can create meaningful differentiation apart from the product itself. However, most brands struggle to find a relevant point of difference and resort to meaningless innovation and sales promotion. What passes for innovation is often mere trivialities: new flavors, scents, colors, and packaging options result in more choice but dilute equity. As expounded by Schwartz in *The Paradox of Choice*, although people believe they want choice, consumers confronted with too much choice tend to avoid making a decision.[6] When innovation simply adds complexity to consumer choices, it is far better to create intangible value as a means of differentiating a brand from its competition.

Even if a product-based innovation proves successful in stimulating consumer demand, it can be readily copied and matched by competitors, and thus, in developed markets, we are faced with crowded shelves full of comparable products. With few substantive reasons on which to base their choices, and lacking any other meaningful differentiation, people often resort to comparing brands on the basis of price. As a result, consumers in developed economies tend to be less brand-loyal and more price-sensitive than those in developing economies. According to BrandZ, half as many people in countries such as the United States, Germany, and the United Kingdom are defined as brand-loyal compared to those living in countries such as China, India, and Brazil, and people in developing economies are more likely to say that it is very important to get the right brand (55 percent versus 41 percent in developed markets).

███ EXPLICIT AND IMPLICIT MEANING

In his book *Brand Meaning*, Mark Batey explores the concept of brand meaning from a number of different angles and draws an important distinction between explicit and implicit brand meaning.[7] The *explicit* or *primary brand meaning* is the meaning that consumers can readily attach to a brand in the course of everyday discussion. Coca-Cola is a refreshing drink. Rolls-Royce cars are for the wealthy. Gillette makes the best razors. But beneath that simple, shared understanding of the brand is a further dimension of meaning: the implicit.

Implicit meaning is the domain of sociocultural and symbolic meaning. It is less evident than the explicit meaning and often harder for people to articulate. As an example, earlier I mentioned "identity" brands. People will easily be able to tell you that Billabong is for surfers and Vans is for skateboarders, but these "tribes" are implicitly linked to a host of different associations that are less easy to pin down but nonetheless real: a different mindset, different landscapes, different music. There is no reason why the two brands cannot market similar products—both sell sandals, for instance—but the rationale for why the brand would sell that product needs to fit with people's understanding of the brand. Actions that go against consumers' implicit understanding of a brand can be more subtle but just as damaging in the long run as trying to change the brand's explicit meaning.

Batey also notes that brands enlist specific symbols to add to their meaning. Symbols like the Marlboro cowboy, the arrow contained in the FedEx logo, and the Harley-Davidson eagle stimulate the imagination and, through the power of suggestion and association, connect the brand to meaningful ideas and values. The square, squat shape of the Jack Daniel's Tennessee Whiskey bottle not only distinguishes it from other liquors but also implies a strength consistent with the brand's shared cultural meaning of independence. Powerful visual cues such as these confer major advantages, particularly when it comes to advertising. Advertising icons such as the Pillsbury Doughboy, the GEICO gecko, and Aleksandr Meerkat (of the U.K. brand ComparetheMarket.com) make their brands easily recognizable. While an advertising icon does not necessarily confer iconic status on a brand, it does at least ensure that the impression conveyed by the advertising is linked to the right brand in people's minds.

NOT EVERYONE FINDS VALUE IN BRANDS

The proportion of people who value brands varies from one category to another and from one country to another for the same category. In any category, some people are inherently brand-loyal, and some are price shoppers. The proportion who value brands as tools to make decisions about what to buy depend on a number of factors in making their decisions: risk, satisfaction with the available options, and perceived substitutability. They believe that there are meaningful differences among brands. By contrast, price shoppers place more value on getting a good deal than buying a specific brand. The group of people to whom you are trying to sell makes a big difference in regard to the strategies and tactics employed.

People are willing to pay a premium for a brand if they feel they are getting something from buying that brand that they cannot get elsewhere. A premium brand needs to justify its premium through meaningful differentiation. This differentiation can be tangible or intangible, obvious or trivial, but in order to create the necessary differentiation, premium brands must invest in innovation and marketing communication. These brands usually make a good margin before subtracting research and development (R&D) and marketing costs and so can afford to spend more than other brands.

Premium brands only need to be different enough to command their existing price point or fend off potential competitors. Differentiation can incur diminishing returns if people have to pay a substantial premium that is out of line with their ability to detect or appreciate what is different.

For example, in the United States, Tide detergent dominates its category but has struggled to justify a price point that was over 50 percent higher than that of its competitors during the Great Recession, when value-oriented brands like Gain were improving their market share. In part, the problem was that Tide's functional differentiation had outstripped consumers' ability to appreciate it. Efforts to remedy the situation led to the development and launch of Tide Pods all-in-one tablets in 2012. *Euromonitor* reports that the revolutionary three-chamber design, which includes a detergent, stain fighter, and brightener, took eight years, 75 technical resources, 6,000 consumers, and over 450 packaging and product designs to develop.[8] In spite of a delayed launch and simultaneous launches of competitive products, Tide Pods has helped Tide grow 9 percent in an otherwise flat market.[9]

Premium brands would do well to stop loyal shoppers from thinking about price, since those people might be horrified if they realized how much they were really paying. Loyal shoppers' behavior is often habitual and unconsidered. (This is not just something that applies to fast-moving consumer goods brands. Until recently, I had owned six Audis in a row. Once, instead of comparison shopping, I simply phoned up the dealer and asked for a black one.) When on autopilot, loyal shoppers tend not to notice price and often underestimate how much they pay for a brand. For premium brands, replenishment and loyalty schemes are a common way of ensuring behavioral loyalty, but creating strong attitudinal loyalty is even better, as it achieves the same effect for less investment.

A value brand needs to maintain the perception that it is an acceptable choice at a lower price than other brands. Suave shampoo in the United States is a classic example of this strategy, presenting itself as just as good as other brands, but at a lower price. True value brands might only compete on price, using discounts and coupons to maintain their position. Other "fair price" brands seek to convey a point of differentiation to make the brand an even more attractive option for consumers. However, that differentiation needs to be supported, either through R&D or through communications. This is tough to do economically when profit margins are thin, as they often are for value brands. In many cases, brand personality becomes the differentiating characteristic rather than functional differentiation.

A Winning Combination for Skoda: A Good Brand at a Good Price

My brother drives a Skoda Fabia. I drive a Volkswagen R32. Both cars are a similar size and do not look that different from each other. If it were not for my ridiculous addiction to speed, a need for four-wheel drive, and the fact that Skodas are not available in the United States, I might well drive a Skoda too. After all, they are reputed to be good vehicles that offer excellent value for the money.

The combination of good functionality and good price has helped Skoda not only ride out but also grow during the Great Recession. A 2012 article in the *Financial Times* reports that the brand's worldwide sales surged 15 percent to 879,200 units in 2011.[10]

It is tempting to suggest that Skoda is just another value brand that has done well during the recession and to predict that its rise will falter when the economy finally recovers, but I think that conclusion would be a mistake. To my mind, Skoda has done a great job of creating a good product and selling it at a fair price. The management team understands that they are a volume player and has tailored the business model and brand positioning accordingly. For instance, Skoda sticks to tried and tested model formats and only introduces new technology when the cost is right. This business and brand model is somewhat different from the usual premium brand model that most people have in mind when they think about brands.

Skoda has done a good job of creating an offer that appeals to the practical and price-sensitive segments of the automotive buyer market. However, according to BrandZ, Skoda is rarely seen as a highly desirable brand outside of its Czech homeland, and based on the overall data, you might have a hard time believing that it is a growing and healthy brand. But brands like Skoda focus on being meaningful and salient *enough*, and they don't need to be different from their competitors or set the trends because they're not asking you to pay a premium for their product.

While its overall equity scores are relatively weak, Skoda's character profile is revealing: it has a reputation for being trustworthy, straightforward, and different. This profile represents an intriguing contrast to its big brother Volkswagen, which is seen as desirable, in control, and trustworthy. The biggest difference between the two is that, among those that have an opinion about each brand, far more people think that Skoda cars are sold at a fair price than are Volkswagen cars (and most other brands, for that matter). The balance between reasonable functionality, positive personality, and fair price means that, in most countries, Skoda is positioned as a good value.

One last thought: being a volume player does not mean a brand is destined to have low margins. The *Financial Times* reported that Skoda earned a five percent operating profit in 2010, a very healthy margin in the highly competitive automotive category. This success makes Skoda one of the chief contributors to the profits of its parent company, Volkswagen, along with Audi and the truck maker Scania.

DIFFERENT PRICE POINTS, DIFFERENT STRATEGIES

A well-differentiated brand *needs* to charge a higher price than its competition in order to recoup the production costs, customer support, creative development, media budget, and so on, that help create its differentiation. With a few notable exceptions, however, brand leaders charge only slightly more than the average price in the category. This is in part because they have extended their reach to more price-sensitive buyers who don't care about the category—that is, people who do not value differentiation in the category. Strong brands typically brand down, supporting premium-priced lines and improving accessibility by producing cheaper versions that require the consumer to give up something; for example, Audi has models that range in price from well over $100,000 to below $30,000 and will even offer different price points for different product configurations within a vehicle model.

How meaningful and different you need to make your brand depends on what price point you need to support in order to fulfill demand profitably. Even high-quality brands need to justify their price point when there are "good enough" alternatives available. In the next chapter, we will "operationalize" the concept of meaningful difference and turn it into metrics that can be linked to how much demand a brand can create, the price premium it can support, and how likely it is to grow market share over time.

Chapter 4

Linking Meaningful Difference to Financial Outcomes

The only reason to develop a brand is if the profit attributable to selling the branded good is higher than the profit that can be made by selling an equivalent non-branded good. Brands create that extra value by influencing the decisions of key stakeholders, not the least of whom is the end customer or consumer. By generating and securing additional consumer demand for the product and reducing the risk associated with future earnings, brands can have a powerful influence on revenue growth and the bottom line.

A strong brand will increase revenues and improve margins in three important ways:

1. By increasing consumers' predisposition to buy the branded product (generating current demand)
2. By increasing consumers' willingness to pay a premium for the branded product relative to alternatives (creating the opportunity to recover a higher margin than otherwise)
3. By increasing the likelihood that consumers will continue buying the product in the future (securing future demand)

To meet any of these goals, the brand needs to influence consumers' purchasing decisions directly. The other benefits of developing a strong brand that have financial ramifications include increasing consumers' propensity to buy additional products from the same brand, creating anticipatory demand in new markets, and improving organizational efficiency.

The influence that a brand exerts on purchase behavior stems from the motivational power of the impressions—both ideas and emotions—that are associated with the brand in consumers' minds. Strong brands predispose people to buy them before they enter the shopping cycle. However, that predisposition is not the only influence exerted on the purchase decision.

It is the interaction between predisposition and in-market activation that determines whether or not someone buys the brand.

PREDISPOSITION

Consistent with the findings from academic research, new research from Millward Brown finds that people are predisposed to buy brands that they find to be meaningful, different, and salient. To varying degrees, these three qualities determine how likely people are to choose the brand, pay a premium for it, and stick with it in future.

Meaningful

The brand must stand for something that people find to be more meaningful than the alternatives. This advantage is rooted in how well the brand is perceived to meet people's functional needs and the strength of the emotional attachment people feel to the brand.

Different

A brand only needs to be marginally different from its competitors in order to be chosen over close alternatives, but our research finds that being seen to set category trends and being viewed as different from other brands in the category improves a brand's ability to command a price premium.

Salient

Brands that come to mind quickly and easily when people are shopping have an immediate advantage over others. They are the easy choice. And if the brand comes to mind with positive and motivating associations attached, then it has an even greater advantage.

If a brand is more meaningful, different, and salient than its competition, then it will likely command the highest-value market share among its target audience, provided it can activate that predisposition effectively. This new understanding has now been incorporated into BrandDynamics™, our brand equity measurement system.

ACTIVATION

Ultimately, any sale is determined by the interaction between predisposition and what happens at the point of purchase. Some people may seek out a specific brand and be willing to pay whatever is asked to get it. Others will end up buying a brand they have never even heard of simply because it is easily available or appears to be a good deal.

Whether or not someone's predisposition turns into a purchase depends on the interaction of that predisposition with activation influences at the point of purchase. These influences help stimulate an individual's predisposition to buy a brand, but the absence of positive activation can create a barrier to purchase. There are two primary influences: findability and affordability.

Findability

To facilitate purchase, your brand needs to be available anywhere people might need it or expect to find it. There is little point in creating a strong predisposition to buy your brand if consumers cannot easily find and purchase it in the form they desire. The exception to this general rule is luxury brands, which tightly control distribution in order to present an image of exclusivity. In the short term, this strategy undermines volume sales, but in the long term, it promotes desirability and allows the brand to command a price premium.

Most companies recognize the importance of making sure their brand is well distributed, but findability also includes visibility—the ease with which people recognize your brand. As noted in Chapter 2, speed of recognition is an important factor in determining impulse and habitual purchases. The ease with which people can see your brand in the store is critical. However, visibility plays a key role in fostering success for any brand. If your brand is easily recognizable, then simply seeing it on the street or online may trigger positive associations in a person's mind, ensuring that the brand remains salient. We will return to findability later in Chapter 10.

Affordability

Generally speaking, the lower a brand is priced relative to the competition, the more people will buy it. Given two equally desirable and visible brands, people are likely to choose the cheaper option. Coupons, promotional offers, and price discounts are all tools by which marketers try to change perceptions of value at the point of purchase.

This summary, however, masks the incredible complexity of the interaction among the nature of the category, the status of the brand, and the price that it can sustain. One of the most important roles of marketing is to affect perceptions of value and justify a brand's price point, and one of the most value-destroying actions that a brand can take is to use price promotion too often, since it shifts people's reference price downward. In other words, people come to expect a lower price and start to shop only when the brand is on sale. When marketers make pricing decisions, they must weigh the short-term volume opportunities against the long-term profit implications if the brand is to remain financially sustainable. We will consider this subject in more detail in Chapter 13.

MEASURING MEANING, DIFFERENCE, AND SALIENCE

As I noted in Chapter 3, the personal meaning that people find in a brand will be as varied as the people themselves. Just like that lump of green stone has meaning for me but none for you, brand meaning will vary from person to person.

Luckily, however, for the purpose of measuring the strength of a brand, meaning can be summarized using two key variables: how well people believe the brand meets their functional needs and how much they like it. Rather than getting bogged down in the details of why people feel the way they do (useful though that is), for evaluation purposes, all we need to do is measure the strength of these two dimensions in the context of a relevant competitive set.

The same is true for difference. To evaluate a brand's ability to command a price premium, we simply need to know whether it is perceived to be setting the trends or is thought to be different from other brands in the product category; we do not need to dig into the details of where that impression comes from.

Obviously, if we want to change these overall perceptions to improve the competitive standing of the brand, then we are going to need to dig deeper to understand the underlying impressions. I will return to the sorts of underlying factors that are important later, but for now, we will simply focus on the top-level association between people's brand impressions and financial outcomes.

Market researchers have traditionally sought to measure salience—or the speed and ease with which brands come to mind—using a category prompt that asks questions like, "When you think about mobile phones, what brands come to mind?" This approach suffers from certain limitations. Traditional, unaided awareness gives some indication of how readily brands come to mind, but by using the category name as the only prompt, this approach fails to stimulate the thoughts or ideas that consumers are likely to consider at the point of purchase. For this reason, we recommend that people be asked to list three main things that they want from the product category and then use those things as a prompt for unaided awareness. For instance, if the respondent says that a big screen is important for their mobile phone, then we ask, "When you think of a big screen, what's the first brand of mobile phone that comes to mind?"

This approach gives a much more accurate prediction of how likely a brand is to come to mind during the shopping process. At an individual respondent level, the correlation between this approach and volume market share is 0.57, in comparison with the correlation of 0.44 produced by a traditional unaided awareness approach (an improvement of 30 percent). The essential difference between these two tactics is that the traditional question assumes that shoppers search just for a generic category product and fails to recognize that consumers can have very specific requirements about which they will be thinking when they shop. This assumption can prove misleading, particularly in highly segmented categories.

LINKING MEANING, DIFFERENCE, AND SALIENCE TO FINANCIAL OUTCOMES

So what evidence do we have that strong brands—those that are more meaningful, different, and salient—perform better in financial terms? The short answer is: a lot. Strong brands make more money using a variety of approaches, each of which highlights a different aspect of brand strength.

Before I review the evidence, I do need to make one thing clear. Meaningfulness, difference, and salience are the three key dimensions of brand strength that can be mapped to financial outcomes, but in order to do so, they need to be analyzed in different ways and given different weights in the final outcome. The same inputs are used in modeling each, but the models are different depending on whether we are predicting current demand, the ability to justify a price premium, or future potential. The other important point to note is that these models use only attitudinal data to create their predictions. Thus, they reflect consumer predispositions that are based on their existing impressions of the brand and category—not what might happen at a later date. We can extrapolate future brand performance from people's current predispositions, but if something unforeseen happens, that potential may not be realized.

As a result of extensive research, we have developed three metrics that measure a respondent's attitudinal predisposition to engage in three different types of behavior. Each type of behavior has clear financial outcomes.

Power: A Single, Accurate Measure of Demand for the Brand

Power is the one-number measure of attitudinal brand strength. Power is based on how meaningful, different, and salient consumers believe a brand to be and is a measure of how well consumers "bond" with a brand. Power measures people's predisposition to buy a brand and can be projected directly to volume market share (before allowing for the influence of in-market facilitators). When it comes to predicting demand for a brand, being meaningful and salient is more important than being different, with the latter only accounting for about 20 percent of the prediction on average.

Premium: The Relative Price Point that the Brand's Equity Can Support

Other equity models—like many marketers—focus on how brands generate volume sales but rarely consider how much consumers are willing to pay for the brand. Our Premium metric explicitly addresses that deficiency. This metric identifies the degree to which consumers believe the brand is worth paying more for and gives important insight into how to balance volume demand and pricing to best grow a brand's financial value. This is where the importance of being different comes to the fore. For a brand to command a premium, it needs to be seen as both meaningful and different, with salience only accounting for about 10 percent of the prediction on average.

Potential: The Anticipation of Financial Value Growth

Based on extensive analysis comparing attitudinal data to changes in value market share over the course of a year, we have improved our ability to anticipate value market share growth. The Potential metric provides an important understanding of how "sticky" the brand is given the competitive context and indicates how attractive new consumers will find the brand to be. Potential does not preclude the need to reach those new customers and make them aware of what the brand has to offer. Although the calculation of a brand's Potential score is very different from the way we calculate its Premium, being perceived as meaningful and different is important to the promotion of both.

SELLING MORE AND CHARGING MORE

In 2012, Millward Brown conducted a pilot study in order to create and validate our new approach to measuring brand strength based on the concepts of meaningfulness, difference, and salience. As part of that process, we compared attitudinal brand equity with actual purchasing behavior as measured by Shopcom's behavioral data.

Shopcom's Data Co-op™ is a single-source, customer-level purchase database of over 231 million U.S. consumers shopping in over 270 consumer packaged goods, retail, automotive, and entertainment sector categories. We merged attitudinal data from our brand equity studies at a respondent level in order to compare Power and Premium scores to actual shopping behavior. The analysis included 65 different brands and 1,600 respondents.

The first important finding is that people who are predicted to have high Power scores (top 25 percent) based on their attitudes are far more likely to buy the brand than those with lower scores (Table 4.1).

What makes this analysis particularly compelling is it does not account for activation influences. The relationship is based purely on respondent predisposition and actual purchasing behavior without accounting for the presence of coupons, out-of-stock merchandise, in-store displays, promotional offers, or price discounts. While people may switch between brands over time, there is a good likelihood that they will buy a brand that they find meaningful, different, and salient. From the summary given previously, it is apparent that those buyers with the highest 25 percent of Power scores account for a disproportionate amount of the brand's volume, buying more than the remaining 75 percent of people combined.

Table 4.1 The power score has a positive relationship with volume sales

Power Score	Volume Bought (Indexed to average of 100)
Low	65
Medium	100
High (top 25%)	324

Table 4.2 The premium score has a positive relationship with price paid

Premium Score	Price Paid (Indexed to average of 100)
Low	93
Medium	100
High (top 25%)	106

Perhaps just as important is the fact that people with high Premium scores (top 25 percent) typically pay more for their purchases (Table 4.2).

This finding is particularly notable given that the in-store price is determined by the brand owner in conjunction with the retail outlet and thus might bear no relationship to the consumer's willingness to pay the price asked. This suggests that consumers who believe a brand to be meaningfully different are willing to pay a higher price for it on average, making them far more valuable to the brand's financial success than the typical shopper. Buyers with the top 25 percent of Premium scores account for almost half of a brand's value market share.

▄▄▄▄ SUCCESSFUL ACTIVATION MATTERS

As I noted previously, the brand equity metrics of Power and Premium are measures of predisposition, and whether or not that predisposition carries through to purchase depends on how well the demand is activated at the point of purchase. This distinction is important because in the quest for growth, activation issues represent the low-hanging fruit. Fixing activation problems should result in immediate returns.

Without going into too much detail, let's briefly compare how well three brands in the U.S. hair care product category convert predisposition into purchase. I have deliberately chosen two brands that lie at either end of the price spectrum in order to highlight the differences between them. The first is Pantene. As a premium brand, Pantene creates a strong predisposition among consumers. It is more likely to be perceived as meaningful, different, and salient than the average brand (Figure 4.1). Its Power score

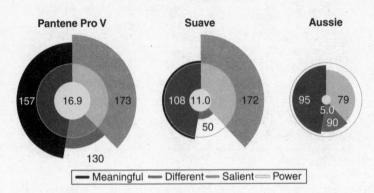

Figure 4.1 Summary comparisons of meaning, difference, and salience

Source: Proprietary research data based on recent purchasers of shampoo.

of 16.9 percent indicates that this would be the brand's market share in the absence of activation influences. The second brand, Suave, is a well-known value brand that creates a good predisposition among consumers but relies far more on salience and far less on differentiation to achieve its Power score than Pantene. The third brand, Aussie, is a less-well-defined brand that lacks salience.

Pantene boosts its actual purchases from the 16.9 percent market share indicated by its Power score to 19.4 percent of category purchases through good in-store activation (Table 4.3). The majority of Pantene's volume sales comes from people predisposed to buy that brand, but a substantial proportion of the predisposed people in this case study ended up buying another brand, representing an opportunity to boost volume market share by a further 5.5 percent. Suave also boosts its volume market share through in-store activation, but it is far more vulnerable than Pantene. Less than half of its volume comes from people predisposed to buy the brand. Aussie fails to activate demand at the point of purchase and so loses sales to competitive brands. Its share of sales is lower than its Power score.

Compared to Suave, Pantene buyers are more likely to be predisposed to choose the brand, so Pantene's existing volume is more secure than Suave's. Suave is far more reliant on activation—particularly offers of good deals and promotions—to generate its market share (Figure 4.2).

Figure 4.2 makes it clear that a failure to keep Suave's price point low and continue to offer price promotions will most likely result in a sharp decline in volume. Both these activation influences add significantly to the brand's total volume market share. By comparison to Suave, Pantene benefits more from distinctive packaging that helps it stand out and Aussie is less likely to be perceived as easy to find. Both Pantene and Aussie have a greater opportunity to boost sales than does Suave (Table 4.3), provided they can identify how to ensure that more of the buyers predisposed to choose them end up doing so.

Table 4.3 Decomposition of volume share differs between brands

% of category volume defined as...	Pantene	Suave	Aussie
Power (Share of Predisposition)	16.9%	11%	5%
Total share of volume	19.4%	18.3%	3.9%
-Secure	11.4%	7.7%	1.5%
-At Risk	8%	10.6%	2.4%
Opportunity	5.5%	3.3%	3.5%

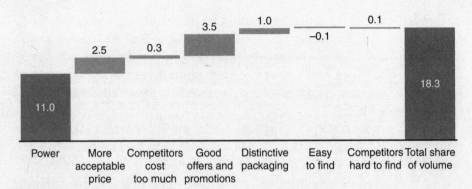

Figure 4.2 Suave benefits from strong price activation

Source: Millward Brown proprietary research data based on recent purchasers of shampoo.

CAPITALIZING ON THE POTENTIAL FOR FUTURE GROWTH

Bertrand Russell is reported to have said, "The demand for certainty is one which is natural to man, but is nevertheless an intellectual vice."

Marketers suffer from this vice no less than others. Unfortunately, many of them assume that the role of market research is to provide certainty, whereas often all it can sensibly provide is the probability that something will happen or not.

This is particularly true when it comes to forecasting the likelihood of future market share growth. No matter how meaningfully different a brand may be, it will not grow unless the marketer manages to make new people aware of what the brand has to offer. Even when sufficient resources are invested to extend awareness of the brand's promise, market share may still decline if a competitor finds a way to disrupt the category or competitors respond with aggressive price promotion. All we can establish from a survey conducted at a single point in time is that a brand possesses the potential to grow based on consumer attitudes at that time.

That said, our measure of Potential does, on average, anticipate a brand's ability to grow value market share in the year following a survey. Brands that people are more likely to believe are meaningful and different are also those that are more likely to grow. Table 4.4 shows that brands with high Potential scores have far higher growth prospects than the average brand, irrespective of product category.

To drive home this point about probability, let me just point out that 47 percent of the high-Potential brands saw a growth in market share, but

Table 4.4 The potential score has a positive relationship with value share growth

Potential	Probability of value share growth 12 months ahead (Indexed to average of 100)
Low	41
Medium	100
High (top 25%)	211

the market share of 25 percent declined in the absence of continued marketing support or in the face of successful competitive action. Conversely, 52 percent of low-Potential brands saw a decline in market share, but 19 percent saw growth as a result of a successful turnaround or weakened competition.

SHARE OF VOICE MATTERS

The implication of what I have said here is that even high-Potential brands need to continue to advertise in order to grow their market share. There is good evidence to support this claim. You not only need to ensure that your brand is meaningfully different from the competition, but you also have to ensure that more people know about it and existing users are reminded of it. You need to amplify what the brand stands for.

Share of voice is a measure of media exposure and refers to a brand's advertising weight of spend as a percentage of total spend for its category. The connection between share of market (SOM) and share of voice (SOV) has been proven in several studies.[1] The higher your SOV compared to your actual market share, the more likely your brand is to grow its market share in the subsequent year (this is often referred to as ESOV, or excess share of voice). So if you increase your marketing investment at a time when competitors are reducing theirs, then you should substantially increase the salience of your brand. This could help you establish an advantage that can be maintained in subsequent years.

Importantly, the studies mentioned previously produce an overall rule of thumb that for every ten points a brand's SOV exceeds its SOM, that brand can expect to gain about one point of market share in the following year. There is, however, variation in that general average depending on product category and brand size. Durables and services tend to benefit far more from increases in SOV than do consumer packaged goods, and large brands tend to gain more than small ones.

SPEND MORE AGAINST A MEANINGFULLY DIFFERENT BRAND

A more recent analysis conducted for this book using attitudinal data from BrandZ compares the share of claimed media awareness with the projected SOM for 1,110 brands measured in both 2011 and 2012. While not using actual market share data, the analysis has the advantage of encompassing media awareness created by the whole media mix, not just measured media like TV. Once again, we find a similar rule of thumb that for every 10 percent growth in perceived media pressure, the predicted market share grows by just over 1 percent.

Using BrandZ data for our analysis offers the advantage of allowing us to examine whether or not brand strength makes a difference in how well

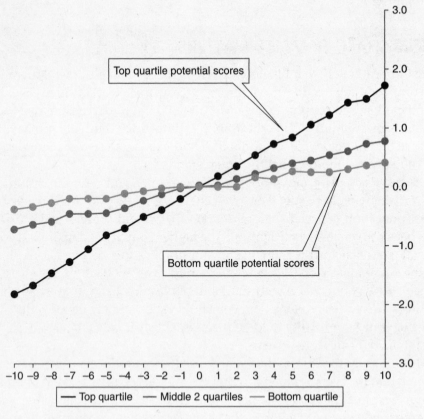

Figure 4.3 High potential brands are more responsive to advertising investment

Source: Millward Brown analysis of Brandz 2011 and 2012.

the brand responds to ESOV. What we have found across a wide array of categories and countries is that brands with strong Potential—those that are poised to grow—are the ones that benefit most an SOV that is higher than their SOM (Figure 4.3). High-Potential brands are more elastic to advertising support than low-Potential brands.

The discerning reader will have noticed that high elasticity has both an upside and a downside. High-Potential brands grow more strongly in response to ESOV, but they also decline more strongly in its absence. In part, this is a reflection of the fact that low ESOV suggests that other brands in the category have something to boast about and are willing to spend to support the brand.

In addition, there are other factors at work apart from advertising and consumer perceptions. Some brands are so meaningfully different from the competition (the Apple iPhone being the most notable) that they tend to grow even more than the high-Potential brand relationship would suggest. Others, like BlackBerry, grow far less. In 2010, BlackBerry was still perceived as meaningful and different from the competition, but its status declined dramatically between 2010 and 2012 as consumer expectations were undermined by service problems and a dearth of exciting new products. Another brand that does less well than its media awareness might imply is Nokia. There is no substitute for a meaningfully different product, and no amount of advertising spend will make up for weak products. In the face of strong competition from the iPhone and the Samsung Galaxy S, BlackBerry and Nokia had nowhere to go but down.

▌ HIGHER REVENUES AND PROFITS

The Power and Premium validation analyses cited previously were limited to consumer packaged goods because they were conducted at a very granular level of individual shopper behavior. In practice, however, these metrics can be created for any brand, and the Potential validation does include durables and services, because we are looking at value market share change. But although we have shown that these metrics do predict changes in behavioral metrics, how do they relate to overall revenue and profits? To demonstrate this relationship, we can map the measures of Power and Potential to overall revenue and profit performance for brands that use the same name as the listed corporation that owns them, such as McDonald's or Vodafone.

In most cases, there is a strong correlation between revenues and brand strength. The respondent-level analysis demonstrates that we should

Table 4.5 Average operating profit reported as a percent of revenues is higher for strong brands

	Low Power	High Power
High Potential	13.6%	15.8%
Low Potential	12.1%	13.2%

expect brands with high aggregate Power scores to generate more demand than weaker ones, and this is what we find. Brands with weak Power and Potential scores tend to underperform the category average and deliver weak revenues, and strong brands tend to overperform. But the really compelling finding is produced when we look at the profits reported by the companies in our analysis.

Table 4.5 presents a simple summary of the data, dividing brands into above- and below-average Power and Potential in terms of operating profit reported as a percentage of revenues across a variety of corporate brands (the data has been normalized so that no one brand has an undue influence on the average). The analysis is based on company annual reports from 49 car, fast food, and communications provider brands and data from our pilot study.

The brands that command the strongest predisposition to buy, as defined by Power and Potential, are the ones that return the highest average operating profit as a percentage of revenues. Brands that score high on both Power and Potential return a profit margin that is 3.7 percentage points (31 percent) higher than that of brands that scored low on both measures. This is the additional return created by a strong brand. Based on the average revenues across the 49 brands, the difference equates to over $387 million a year.

While Power's correlation with brand size means that other influences, like positive returns to scale, might account for the stronger profit percentage on that axis, Potential is not correlated with brand size. On average, high-Power brands with above-average Potential in our dataset return a profit margin 2.6 percentage points (20 percent) higher than that of brands with below-average Potential. For low-Power brands, the difference is less, but high-Potential brands return profit levels 1.5 percentage points (over 12 percent) higher than that returned by low-Potential brands.

By creating the perception among individual consumers that they are meaningful, different, and salient, strong brands:

- Capture three times more volume sales than the average brand
- Command a price premium 6 percent above the average
- Are two times more likely to grow value share over the next twelve months

The end result is that strong brands generate higher profits. This analysis brings us back full circle to the finding that strong brands tend to outperform the S&P 500. The stock market rewards companies that exhibit profit and strong growth potential, and meaningful, different, and salient brands deliver what the stock market wants.

PART II

CREATING A MEANINGFULLY DIFFERENT BRAND

Brands that are meaningful, different, and salient are stronger and more valuable than brands that are not. Unfortunately, strong brands are few and far between. Even if you are lucky enough to work with a meaningfully different brand, there is still probably room for improvement—new ways by which you can generate better financial returns from your brand. Meaningful difference can boost the financial return from any brand, but first you must define what makes your brand meaningfully different to its consumers.

This section of the book will focus on the four components of a meaningfully different brand: purpose, delivery, resonance, and difference.

IS THE UNDERSTANDING OF YOUR BRAND SHARED?

At this point, I suspect that some readers will be preparing to skip to the amplification section in the confident belief that they know what makes their brand meaningfully different. Maybe this difference is written down somewhere. Perhaps the definition is embodied in a brand key like the ones used by Unilever, Kellogg, and Diageo.

And maybe you do know what makes your brand meaningfully different. But do your colleagues? Do the sales and marketing teams understand the brand in the same way? What about the people in customer service? And what about the tellers, concierges, servers, attendants, or salespeople? Does the behavior of the people talking to your customers embody what makes your brand meaningfully different? And what about your design, PR, ad, digital and sales promotion agency teams? Do they get it? Do the brand's collateral and marketing communications convey a unified impression that is meaningfully different?

If you ask them, you may find that people simply parrot whatever is written in the brand's strategy document. Superficially, everyone may appear to know what makes the brand meaningful and different, but if you dig a little deeper and ask them to explain the brand's purpose and promise in their own words, then you may find cracks starting to appear. In our ValueDrivers workshop, people working on the same brand often have very different understandings of what that brand stands for. And a lack of clarity not only breeds poor internal alignment but also makes a very shaky foundation on which to build a brand's promise to its consumers.

So go on, read about how to define your brand's meaningfully different experience. I promise it will be time well spent—because once you really know what your brand stands for, implementation becomes a whole lot easier.

Chapter 5

Clarity of Purpose

Charles Dodgson is best known for writing *Alice's Adventures in Wonderland* under the pen name Lewis Carroll. Dodgson is often credited with saying, "If you don't know where you are going, any road will get you there." In spite of its source, this statement is pragmatic, not fantastical, and it is just as true for brands as it is for people. A brand must have a purpose. It must make some difference in people's lives. Otherwise, why are people going to pay to use it? Purpose is the foundation on which strong brands are built. To justify its existence in today's complex and crowded categories, a brand needs a purpose that sets it apart from others. What does this brand offer that others do not?

WHY DOES YOUR BRAND EXIST?

At minimum, a brand must have some basic functional purpose; it must provide something consumers want or need. A brand can satisfy emotional needs along with practical ones when it triggers associations related to things like self-worth, pride, caring, or security. A brand's purpose can also be informed by the brand's story or heritage. As Diageo's Mark Murray notes: "Exploring the brand's roots is always a worthwhile exercise. What made it interesting to people in the first place can tell you a lot about a brand."[1]

Some brands are able to elevate their purpose to the level of an ideal that goes far beyond the functional delivery of the product to address higher-order needs such as self-fulfillment, identity, affiliation, and societal or environmental good. Let's consider the different ways in which a brand might seek to make a difference in people's lives.

DO THE SAME JOB, ONLY BETTER

The origins of Dyson vacuum cleaners are well known. Frustrated with a premium vacuum cleaner that failed to live up to his expectations, James

Dyson set out to build a better one. It took fifteen years and over five thousand prototypes, but in May 1993, he launched the DC01, billed as the first vacuum cleaner that would never lose suction. The technology was not just better than that of existing vacuums—it was *visibly* better. People were fascinated to see the dirt whizzing around in the transparent container. The sight was proof that the product was working and confirmed how different it was from the traditional bag technology. The new cleaner was an immediate hit, and in 2011, the company generated over $1.5 billion in revenues. What is perhaps less well known is that existing vacuum manufacturers had previously passed up the opportunity to acquire the revolutionary cyclone technology (a decision they lived to regret), forcing Dyson to fund his own start-up and helping him become a billionaire.

The Dyson company continues to reflect James Dyson's commitment to radical problem-solving. Dyson seeks to improve electrical appliances through better engineering and design. If there is a standard and accepted way to achieve a goal, then they aim to find a new and better way. We might sum up the company purpose as "better by design." The results speak for themselves. The tiny DC12 vacuum cleaner, specifically designed for Japan, a country where space is at a premium, boasts a power-to-weight ratio five times greater than that of a Formula One racing car and gained 20 percent of the market within three months of its launch, becoming the market leader in 2006.[2] That same year, Dyson introduced the AirBlade hand dryer, which uses a jet of air to squeegee water off hands and consumes far less electricity than conventional dryers. In 2010, the company introduced the Air Multiplier bladeless fan. Dyson products typically retail at far higher prices than the alternatives, but James Dyson asserts that people are willing to pay more for a demonstrably better product.

SEE A NEED, FILL A NEED

Coway is a South Korean brand whose purpose is to make the world in which people live a happier, healthier place. Coway is notable not just for its purpose but for the innovative way in which it has made access to its products more affordable and extended its business from home electronics to services.

Coway was created in 1989 and initially focused on producing electronic water filtration products for the home. Today the Seoul-based company is a market leader in water and air purifiers, and its portfolio also includes water softeners, vacuum cleaners, and more. In 2010, Coway generated sales of $1.43 billion, marking its twelfth consecutive year of record sales.[3] The company not only sells home electronics focused on a healthy environment but also rents them and, in addition, offers unique home care services.

Following the Asian financial crisis of 1997, South Korea suffered an economic slowdown, but, as is so often the case, hard times stimulated a creative response. In April 1998, under the leadership of the recently appointed CEO Yong-Sun Park, Coway started a rental business. The new service offered consumers the chance to rent water or air purifiers and reduced the initial high cost of purchasing one. This move was a first for a Korean brand, and by 2012, Coway's rental service had attracted over 5.5 million users and accounted for the majority of the brand's revenues. This case is a great example of how to make a brand more affordable, extend access to people who could otherwise not afford it, and grow revenues as a result.

As part of the rental service, a trained purifier expert visits customers on a regular basis, changing used filters and ensuring that products are in optimal condition—free of charge. This "Cody System," along with the company's "Heart Service," which guarantees service within three hours if needed, has helped the brand establish an important presence in many Korean homes. Coway has also crowdsourced ideas for future development. In 2012, staff and customers generated 12,000 ideas, 30 percent of which were used to improve customer service.

Building on its strong reputation, in late 2011, Coway launched a new product line: mattresses. People spend a large part of their lives asleep, but good-quality mattresses are costly items beyond the reach of many household budgets. So, like Coway's home electronics products, the mattresses can be rented as well as bought, and each comes with its own "Home Care Doctor Service." This specialized cleaning service offers professional checkup and intensive hygienic care for mattresses and has now been extended to carpets and other household items. Coway's 2012 third-quarter earnings report noted that the new mattress business had over 100,000 rental accounts and had achieved a seven percent market share by sales.

While Coway's business performance has suffered during the recent economic downturn in Korea, the brand remains successful and profitable and has expanded into Thailand, China, and the United States. As an indicator of the value Coway has created, in January 2013, the struggling Woongjin Holdings Co. Ltd. sold its entire 28.37 percent stake in Coway for $1.1 billion to private equity fund MBK Partners. Current Coway CEO Joon-Kee Hong has stated: "As the wellness boom sweeps the world as a mega trend, Coway is preparing itself to emerge as a global well-being company, capitalizing on our excellence in R&D, design and overseas businesses which help shape our pioneering thinking such as 'rental business' and 'CODY system.' We are confident that in the near future, our products will be able to serve you wherever you are."[4] If this statement proves true, then the investment by MBK Partners could prove a very profitable one.

Clearly, the idea of changing people's lives for the better has not limited Coway's growth potential but has instead empowered it. Rather than being defined by a specific product category, the brand has been able to extend from a manufacturing base to selling services under its "We Bring Wellness" umbrella.

BEYOND PURPOSE TO THE PURSUIT OF AN IDEAL

"Maximum growth and high ideals are not incompatible. They're inseparable."

So begins the introductory chapter to Jim Stengel's book *Grow: How Ideals Power Growth and Profit at the World's Greatest Companies.*[5] It is a provocative statement, particularly for those challenged to grow their business profitably during tough economic times, when it is tempting to do whatever it takes to keep a business afloat. But, as the example of Coway demonstrates, pursuing an ideal can empower business success.

To my mind, ideals are a higher-order purpose. Your brand may not need ideals in order to be successful, but it certainly needs a purpose. If you can help people achieve their goals better than the alternatives can, then you will already be one step ahead, but people seek more than just a job well done. They seek meaning in their lives. An ideal provides a brand with an inspirational reason for being—one that people can buy into.

A brand ideal is not a mission statement. Mission statements tend to be narrow, business-oriented phrases such as "Be the best company in our industry" or "Be the most innovative company." Mission statements tend to be self-serving and therefore limiting. Ideals, being outwardly focused, extend beyond the company's financial interests.

Stengel refers to ideals as the 400 percent advantage. Why? Because an investment in the Stengel 50, the top 50 businesses found in a ten-year study of growth, would have outperformed the S&P 500 by a factor of four over the past decade. The Stengel 50 were identified from an analysis conducted by Stengel in conjunction with Millward Brown Optimor, which conducted a similar study for Jim Stengel when he was the global marketing officer at Procter & Gamble. The second Stengel study improved on its predecessor in that it spanned a whole decade rather than five years and took into account both financial value growth and how well the brands grew their attitudinal bond with customers. Representing a wide variety of product and service categories, the Stengel 50 are united by one common factor: they operate in harmony with their ideals. For these companies, ideals provide a "North Star," a compass bearing by which they can steer through good times and bad.

Ideals help create clarity. As Stengel says, "A viable brand ideal cuts through the clutter and clarifies what you and your people stand for and believe."[6] Why does this matter? It matters because belief is a powerful motivator. If someone believes that his or her company is dedicated to improving people's lives, then he or she will give that extra ten percent to make good things happen. In the absence of clarity about what the brand stands for, people are simply doing a job and must find their own motivations for doing it well.

Zappos, the full-price online and telephone shoe retailer, exemplifies the power of ideals to create clarity of purpose and positive motivation. Zappos exists to bring happiness through excellent service. In a world of businesses that do everything to prevent customers from speaking to a live person, Zappos encourages its call center staff to spend time on the phone with customers. Their objective is not just to satisfy customers but to make them happy. There are no scripts, no time limits, and no sales quotas. A seven-hour service call is celebrated, not condemned. The end result? Massively satisfied customers and explosive growth. Oh, and a lot of fans on Facebook—all without offering a single discount.

Of course, ideals alone do not create profitable growth. Zappos would not be successful if it did not have a honed and efficient supply chain and logistics. But a strong ideal helps keep a brand on track and acts as a multiplier to all the other elements of a brand, including marketing communications. As an indicator of its financial value, in 2009 Amazon bought Zappos in exchange for 10 million shares (approximately $880 million), with Zappos employees getting a further $40 million in cash and restricted stock.

Though the main influence of ideals is to create unity and motivation within a brand, a brand's ideal is not necessarily something that needs to be communicated outside the organization. In research conducted to explore this topic, a selection of the Stengel 50 were only slightly more likely than a set of comparable benchmark brands to be credited with trying to make people's lives better (29 percent versus 24 percent). However, in work conducted by Millward Brown's consumer neuroscience practice, the Stengel 50 brands were found to be more empathetic, more ideals-based, and more memorable in terms of what they stand for than their competitors. The end result is that the Stengel 50 brands are more likely to be considered for purchase and have a higher potential to grow than the benchmark brands.

▰ NATURA SEEKS TO CREATE WELLNESS AND WELL-BEING

In 1969, Antônio Luiz da Cunha Seabra founded Natura, the Brazilian cosmetics and personal care company, as a counterpoint to the ethos of the established cosmetics industry. He wanted Natura to make people feel good about themselves by offering them a way to express themselves and connect with others and the environment, all without resorting to the potentially toxic ingredients included in some mainstream brands.

In 1974, the company opted for a direct sales model using consultants, similar to that used by Avon. The direct sales model was well-suited to the Brazilian market, given its lack of a modern retail trade, and offered good job opportunities for women. Today there are 1.4 million Natura consultants.[7] While this business model has undoubtedly helped fuel the brand's success in Latin America, Natura is also lauded for embracing sustainability and the use of natural ingredients in its formulas. For instance, in 2009, Natura partnered with the Water Footprint Network to inventory the company's use of water. Since then, it has implemented several initiatives to reduce water consumption. Natura also seeks to source natural ingredients sustainably and will not produce hair dyes or nail polish, because necessary ingredients may prove harmful to their users.

Today, the company's purpose reflects Seabra's original vision. That purpose is to create and market products and services that promote *"bem estar bem"*—"wellness and wellbeing"—an aim that is infused throughout the culture and operations of the company. In a presentation given in 2012, José Vicente Marino, vice president of commerce at Natura, explained the Natura brand this way: "Above all is it our way of being. The brand is everything we do: the product, the service, the marketing, the ambience. The brand is everything we do *and how we do it*: with purpose, principles, beauty, and commitment."[8]

Statements of this sort might be written off as simply good intentions or hyperbole if it were not for the fact that Natura rigorously defines and measures its progress by and against its purpose. Today Natura aims to transform social and environmental challenges into opportunities for business by espousing the philosophy of the *triple bottom line*. The triple bottom line encompasses the aims of ensuring strong cash flow for good financial results; social impact through the generation of wealth for their consultants; and environmental impact through encouraging the use of refills, sustainable extraction, and the use of recycled and recyclable materials.

I have met with executives from Natura and can personally vouch for the passion created by the corporate commitment to these goals. That commitment has powered the brand's growth. In 2011, net revenues reached $2.6 billion, up nearly nine percent over the prior year; earnings before interest, taxes, depreciation, and amortization (EBITDA) were $700 million, with an EBITDA margin of 25.5 percent. Over $1 billion was distributed among its consultants and supplier communities that same year. In spite of significant growth, the company reduced its greenhouse gas emissions by 3.12 percent and its water consumption by 4.8 percent. Refills now account for 17 percent of sales. In the 2012 Top 50 Most Valuable Latin American Brands rankings, the Natura brand ranked number 15, with a value of $3.3 billion.

Both Natura and Unilever's Dove brand are in the Stengel 50, and you might have noted some similarity between the two in terms of their purpose. At the heart of the Dove brand is its commitment to its Campaign For Real Beauty, which seeks to inspire women to be comfortable with themselves and to have confidence in their own unique beauty. The appeal of the Dove brand has been successfully extended across product categories and countries to achieve significant earnings growth over the last decade. Dove has been a significant brand in Brazil for some time, but in spite of the same degree of familiarity and perceived performance on key category criteria, over four times as many Brazilians who buy body care products claim that the last brand they purchased was Natura, not Dove. Moreover, they are likely to have paid a higher price for their purchases. In Brazil, Natura commands what we call a *justified premium positioning* compared to Dove's good-value positioning.

Why is Dove not stronger in Brazil? Because Natura had already established itself as the brand of inner confidence and wellbeing before Dove arrived in Brazil, and consumers are still convinced that Natura products are good. A large proportion of Brazilian body care consumers agree that Natura is scientifically advanced, keeps skin in better condition, and offers a good range of products, and these functional benefits combine to produce a strong emotional appeal. Seventy-four percent are willing to agree that Natura is a brand they trust, and 65 percent agree that it is a company that is leading the way in the field. Global brands like Dove and Avon may be as well known as Natura, but they simply do not inspire the same degree of passion. Fifty-four percent of people claim to have recommended Natura to a friend, compared to 41 percent who have recommended Avon and 22 percent who have recommended Dove.

MAKING THE UNIMPORTANT IMPORTANT

To what degree does Natura's sustainability commitment drive the brand's appeal? It is certainly well known that Natura espouses a green agenda. In 2010, 64 percent of personal care buyers in Brazil interviewed for BrandZ agreed that the brand behaves in a responsible way toward the environment. And it is true that the environment seems to have more influence on loyalty in this category and country than it does in others. But is that a cause of Natura's strength or the result of it? I would argue that Natura's environmental credentials are not the key driver of the brand's success with its consumers, but a supplementary one. Consumers love the brand for what it does for them, first and foremost. What is far more important is that Natura's commitment to sustainability motivates its employees to do a better job.

This is not to say that sustainability does not offer competitive advantage, but I would argue that by pursuing a sustainability agenda, Natura has added another layer of comparison to the product categories in which it competes. All things being equal, the brand's commitment to sustainability gives consumers an additional reason to choose Natura over the competition. Right now, that reason may be well ahead of the existing consumer mindset, but you cannot judge the importance of an attribute or association simply based on what consumers say today. It is up to the brand's management team to decide whether such an association can make a difference and, if it can, to act. If the judgment is correct and the implementation effective, then the new association will become important as people choose between brands.

Comparison with the United States and United Kingdom using the same analysis finds that trying to be environmentally friendly has far less importance in those countries than it does in Brazil. In the United States, it ranks as the least important influence on purchase, and in the United Kingdom, it ranks fourteenth out of fifteen attributes. To my mind, this suggests that Natura is creating a motivation in Brazil that has not been addressed in the other two countries (although the slightly higher ranking in the United Kingdom may reflect the presence of the Body Shop, which has also espoused a more sustainable positioning). Because Natura satisfies basic category needs as well as or better than the competition, its environmental friendliness then becomes the deciding factor in whether or not someone chooses the brand. But without the functional credentials and trust, Natura would not be the powerhouse it is today.

A SENSE OF PURPOSE IS EASY TO LOSE

People use brands to perform tasks. Theodore Levitt recognized this fact when he famously stated, "People don't want to buy a quarter-inch drill. They want a quarter-inch hole!"[9] Put this way, this concept seems self-evident, but time and again I find that companies have lost sight of why people really buy their brands. Why then do we forget this basic fact so easily? The answer is that we get blinded by our own viewpoint. What matters most to us tends to dominate what we think matters most to our customers. This is not to imply that we are ill-willed or selfish; rather, we are simply blinkered by the immediate tasks we need to perform to be successful: sell more, cut costs, seek efficiencies. Even if you are the founder of a company, the day-to-day needs of a business can drown out your sense of purpose.

A couple of years ago, when developing the ValueDrivers workshop, I spent a couple of days working with the management team of Shackleton Thomas to figure out how to reinvigorate their brand. Founded by Charles Shackleton and Miranda Thomas in 1986, the company makes and sells handcrafted furniture, pottery, and more from its workshops in a 180-year-old mill in Bridgewater, Vermont. Charles designs and oversees the furniture making, and Miranda designs and oversees the pottery making. Like many businesses, ShackletonThomas has had its ups and downs, and 2010 found the company struggling with the effects of the Great Recession on discretionary spending. We needed to find a way to boost revenues, but how could we do so when the company's products were so expensive to make?

The first part of the ValueDrivers workshop focuses on purpose. Why does the brand exist? This proved a tough question for ShackletonThomas's management team to answer. We tackled the issue from a number of different directions and ended up with plenty of good material—too much, actually, because we kept getting lost in the detail. It was clear that Charles and Miranda believe products made by hand possess a character—a soul, if you like—that machine-made products do not. Their aim is not perfection; rather, they aim to display the personality of the craftsman through the physical object, blemishes and all, and they believe that their customers appreciate these same properties. But we could not pin this belief down to a succinct set of words.

Then we had a breakthrough. Suddenly the phrase "a passion for hand-craft" was spoken and captured what the company was all about. It was at this point that Charles jumped up and announced, "That's why we created this company. I wrote it in the manual!" Sure enough, we had rediscovered why Charles and Miranda had decided to go into business for themselves. We had rediscovered the company's original purpose. But in the intervening years, that purpose had been subordinated to the distractions of running the day-to-day business—designing new pieces, negotiating for materials, training craftspeople, making payroll, talking with prospective customers, dealing with the bank—and the need to drive revenues.

Today, in spite of a body blow dealt when Tropical Storm Irene flooded the company's workshops, ShackletonThomas is going strong, with a rejuvenated sense of purpose. The idea of a passion for handcraft is made visible in everything the company does. The workshops have been reconfigured so that visitors can see the company's signature products being made. The displays have been redesigned to tell the story and explain the unique nature of the company's products. No longer will people visit and come away wondering what makes ShackletonThomas different from its machine-focused equivalents. Not everyone will appreciate the difference, and not everyone will be able to afford the products, but at least they will under-stand that each item is, in its own way, unique.

If the founders of a company can lose sight of their original purpose, then how easy is it for the same thing to happen when the founders are no longer present? There is an old adage that suggests "a new broom sweeps clean." When it comes to business, sometimes the broom sweeps away the company's sense of purpose as well.

THE STRUGGLE TO DEFINE PURPOSE

Natura and Hewlett-Packard (HP) share something in common. Both com-panies were founded in a garage: the former in 1969 in São Paulo and the latter in Palo Alto in 1939. Thereafter, the similarity ends. One is successful with a clear sense of purpose, and the other is struggling to reinvent itself. That struggle became very apparent in a recent interview broadcast on national public radio in the United States.

On the same day that HP announced a $9 billion write-off on the acquisi-tion of the U.K. software company Autonomy, HP CEO Meg Whitman joined Kai Ryssdal, host of American Public Media's *Marketplace* show, for a "Corner Office" interview. Having dispensed with the issue of the write-off and discussed the role of innovation in HP's future, Ryssdal asked what one might have thought was a simple question: What is HP? He requested that Whitman cut out the marketing gobbledygook and executive-speak and put the answer into plain English.

Whitman spent nearly a minute describing the two customer bases served by HP—consumers and chief information officers, or senior leaders at HP's value-added resellers—and admitted that this dichotomy "makes HP a little bit more complicated."[10]

After a brief diversion to consider whether HP might be more successful focusing on the enterprise business alone, Ryssdal returned to the question of what HP stands for, stating: "Recognizing that HP does have two cus-tomer bases, I do have to point out that you did not really answer the ques-tion of what HP is."[11]

This time Whitman's answer was somewhat more concise: "HP is the largest information technology company in the world that provides hard-ware, software, and services to organizations of all sizes and has a big con-sumer business as well. What we do is we try to provide enterprise needs better than anyone else and deliver the hardware and software and services that those enterprises need. And put together a series of devices that customers want as well for their printing and personal systems needs."[12]

So what is wrong with this description? For a start, it is too long. If this were the so-called elevator pitch, then we would likely be out of the elevator before it was over. This is not a trivial issue. If you cannot succinctly state what difference your brand makes in the lives of your customers and consumers, then you probably don't really know.

That is the far bigger problem with Whitman's answer. Whitman describes what the company does, not what difference it makes in customers' businesses or their personal lives. She does say that HP tries to serve enterprise needs better than anyone else, but that is a generic statement that any supplier is going to make. What she fails to get across is what makes HP meaningfully different to its customers. In what way will HP empower its customers to run their business, do their jobs, or lead their lives better than another IT company?

Sure, HP is a big, complex organization, but that is all the more reason to have absolute clarity about what its purpose is. Lack of clarity simply leads to poor organizational alignment and poor external communication, as witnessed in this interview. Not surprisingly, Whitman is not alone in failing to have a succinct answer for what her company stands for. A few weeks later, Tim Armstrong, AOL's CEO, failed the same test, although to be fair, on the second go-round he did manage to boil his reply down to, "The best media and technology brand company in the world."[13] Strange, I could have sworn that was Sony...or Google...or even Yahoo!—but not AOL.

I would like to say that this lack of clarity regarding a brand's purpose is unusual, but sadly, it is not. I have talked to senior executives of major brands all over the world, and all too often, they struggle to state what their brand's purpose might be.

More sophisticated brand-led companies know how important it is to clarify what their brand stands for. They define the brand's promise in what is sometimes called the *brand key*. All too often, however, the key opens up a maze that includes not just the brand's purpose but also its functional and emotional benefits, brand personality, the nature of the target audience, and more. Maurice Saatchi's "one-word equity" becomes a cloud of phrases and attributes that serves to hide, rather than reveal, the brand's true nature.

◼︎◼︎◼︎ UNDERSTAND YOUR PURPOSE

A brand's purpose should be clear to everyone in the company. An organization is much more likely to be single-minded in its focus and to speak with one voice when it is permeated by an understanding of the brand's purpose. Implicit in the idea of purpose is the need to know whom the brand is intended to benefit. Who are these people, and what difference is the brand going to make in their lives? For a brand to be perceived as meaningfully different, it must offer something its competitors do not, and that offer must resonate with customers. This is most likely to happen when a business is clear about the purpose of the brand and when the brand delivers the differentiated experience it promises.

In some cases, a leader who is passionately concerned about the brand can elevate its purpose to an ideal, inspiring and commanding others to follow. Ideals exert their strongest influence on the people who work on a brand, but they also have an indirect effect on customers and consumers. Even though the ideal may not be known to the general public, by informing the brand's actions and motivating staff, that ideal will effectively separate the brand from others whose only purpose is to meet a functional need.

Questions to Consider

1. Do all your colleagues and brand stakeholders (media agencies, communications agencies, research agencies, consultants) have a good understanding of the purpose of the brand? Would they describe that purpose in the same way? Do they believe in it? Do they act like they believe in it? You could ask each the same question: "How do you think this brand makes a difference in people's lives?"

2. If there is a lack of clarity regarding the brand's purpose, how will you implement a process to develop a definition of that purpose? A purpose will rarely take root if it is not enthusiastically endorsed by the most senior and involved members of the company. For a unibrand company (and indeed for many big brands in multibrand companies), this will mean the CEO and the main board. Invite people to a working session during which they can each submit their definition anonymously and in advance. Review the definitions and identify similarities and differences between them. Create a new definition, starting with the history of the brand—why and how it came into being—and work from there. Remember, the focus should be your target consumer. How does the brand make a difference in his or her life? Aim for something inspirational, something that can inspire everyone working on the brand.

Chapter 6

Effective Delivery

If purpose is the difference that a brand promises to make to people's lives, then delivery is how well the brand fulfills that promise. How good is the actual brand experience? Good delivery is critical to a brand's long-term success, because meaning originates from the brand experience. Ensuring that the brand's meaningful difference is inherent across all touch points is key to good delivery. The way the brand presents itself to the target audience may shape that experience, but it won't make a bad experience into a good one. If there are weaknesses in the brand experience, then you need to identify them and address them immediately. Strong brands might be forgiven a bad experience, but they cannot count on it.

At one level, delivery comprises the tangible product or service experience. If a brand fails to match up to the general expectations of the category, then it is in trouble from the start (unless it is priced accordingly). If it fails to deliver on specific claims, then people will also be disappointed. But delivery comprises a much broader context than just the product itself. As noted previously, a brand is everything from purpose to calls to action, from customer service to pricing, from warranties to communication. Every single interaction with customers contributes to the brand experience. Defined in this way, delivery is simply how well the company walks its talk.

What the brand stands for—the things that make it meaningfully different from the competition—should be made readily apparent through the brand experience. As noted before, differentiation is most potent when it is intrinsic, that is, based on relevant and tangible advantages and powered by the declared purpose of the brand. Intrinsic differentiation can be accomplished through the look, feel, sound, smell, or taste of a product. The delivery of an outstanding brand experience depends on attention to all of the details: the consistency of the product, the functionality of the brand's website, the clarity of its usage instructions, and so on.

For service businesses, human delivery is the heart of the brand experience. A company culture built around the brand's purpose or ideal will help ensure consistent, high-quality service delivery. However, human factors contribute to the experience of other types of brands as well—for example, the brand representatives encountered at retail outlets, car dealerships, and call centers.

THE SENSORY BRAND EXPERIENCE

In 2004, I worked with Martin Lindstrom to design and implement a research study examining the impact of the sensory experience on brand loyalty. The overall conclusion of the research was that prior sensory experiences, particularly those that created a strong positive emotional response, do affect brand loyalty. The more positive and distinctive sensory impressions a brand generates, the more loyal people are to a brand. Findings from this research are reported in more depth in Chapter 6 of Lindstrom's book *Brand Sense*.[1]

Since the book was published, some elements of sensory marketing have been widely adopted. Scent marketing in particular has become popular. Walk into a hotel today and you are likely to be greeted by its signature scent. In 2006 Starwood Hotels and Resorts announced that the signature scent for its Sheraton Four Points hotels would be that of apple cinnamon pie. Called "Uncomplicated Comfort," the scent joined "White Tea" (Westin Hotels) and "Welcoming Warmth" (Sheraton Hotels) on Starwood's scent roster. There is good evidence to suggest that our sense of smell has a strong influence on our appreciation of our environment and can influence our predisposition to spend more when encountered in-store.

While scent marketing has been widely adopted, the sense of touch may be one of the least considered and most influential motivators of purchase. A paper authored by Joann Peck and Susan B. Shu suggests that "the influence of the sense of touch may create the desire to possess an object."[2] To begin their discussion, Peck and Shu quote a brief dialog from the movie *Star Trek: First Contact*:

> *Captain Jean-Luc Picard*: It's a boyhood fantasy....I must have seen this ship hundreds of times in the Smithsonian, but I was never able to touch it.
>
> *Lieutenant-Commander Data*: Sir, does tactile contact alter your perception?
>
> *Picard*: Oh yes! For humans, touch can connect you to an object in a very personal way.

After reviewing the results of their experiments into the effect of touch on perceptions of ownership and value, Peck and Shu conclude that their research "supports Jean-Luc Picard's claim.... In four studies, we found that touch does connect a person to an object by increasing the feeling of ownership of the object."[3]

The results of their research make intuitive sense to me. Before I succumbed to my desire to buy an iPod Touch, I visited the Apple store twice. Each time, I picked up the iPod Touch and handled it. The experience of touching the device spoke volumes and had more effect on my ultimate decision to purchase one than the advertising I had seen. But, having said that, I think the authors downplay an important finding from their research.

Peck and Shu's results clearly demonstrate that touching a product does increase perceptions of value. But they also demonstrate that when people are unable to touch an object, they value it more highly if they imagine they own it. The magnitude of these effects is comparable, if not biased in favor of imagined ownership. (The average valuation of the objects tested was $3.59 when people could not touch it, compared to $3.34 when they could.) This finding is similar to the rehearsal effect that I described in Chapter 2. I believe that both groups of people—the ones who touched the products and the ones who did not—imagined what it would be like to own them. Therefore, Peck and Shu's research leads me to a more general conclusion than that of the authors. My conclusion would be that encouraging feelings of ownership can have a significant impact on a brand's perceived value.

I believe that the positive sensation of touch is ignored by many businesses. Recently, Gordon Pincott and I were conducting a ValueDrivers workshop for a large, well-known Colombian food brand. During the course of the workshop it became apparent that in order to realize cost efficiencies, the company had changed an important part of the product experience, one that involved the sense of touch and had formed the basis of an almost-universal ritual for the many people who used the brand. The new, cheaper packaging format not only denied the pleasure created by the ritual but on some occasions also resulted in a negative experience for consumers. By seeking to save money, the company had actually undermined the brand experience and potentially weakened consumer loyalty.

A related issue is sometimes referred to as "the death by a thousand cuts," or the all-too-relevant "creeping normalcy." In this scenario, the product specification is reviewed on a regular basis in order to identify costly components and either remove them or replace them with cheaper ones. When consumer testing is conducted (sadly not a common practice these days), the cheaper version wins if the preference for the prior formulation is marginal. Unfortunately, when the same rationale is applied to a series of incremental changes over a period of years, you often end up with a product that is notably inferior to the ones tested at the start of the process.

Many years ago, Cadbury eroded the equity of its Dairy Milk chocolate bar by making the bar thinner. This allowed the brand to look the same and retain the same price but improved the margin on each bar sold. However, the process went too far, and competitor Rowntree Mackintosh (now owned by Nestlé) launched a chunky chocolate bar called Yorkie. Supported by an ad campaign targeted to men, the launch was successful and forced Cadbury to boost the thickness of Cadbury Dairy Milk in response.

Worse than unplanned cost cutting is the planned "dumbing down" of products following a successful launch. This usually occurs when the product specification at launch is deemed too expensive to be sustainable, and so as soon as the product is established, cost is stripped out of the product (or packaging). Strategies like this seem completely futile. If the product specification cannot be sustained at its initial price point, then the chance of long-term success is negligible. Even if consumers do not recognize them on their own, competitors will likely draw attention to any product shortcomings.

PUTTING A VALUE ON THE SENSORY EXPERIENCE OF A BRAND

One of the biggest challenges facing marketers is that it is always easier to make the case for cost reduction than the case for not giving something up. Why? Because you can easily document cost savings, but you cannot easily document opportunity cost. In the case of the Colombian food brand, I am sure that there was a well-reasoned and documented case made that detailed how much money could be saved over a specific time frame by investing in new packaging equipment. What would have been far more difficult to quantify was the cost to consumer satisfaction and loyalty of making the change to packaging that offered an inferior format. The value of what was lost could have been significant.

Part of the research described in *Brand Sense* focused on Dove soap. Consumers referred to Dove's scent as "subtle," with one user saying, "Dove is a pure soap and I smell its cleanness and purity when using it."[4] The results for Dove suggested that the smell of the brand makes users feel relaxed, valued, and calm; it leaves people feeling satisfied. We also found that smell is very valuable to the Dove brand, accounting for $63 million of Dove's annual revenues in the U.S. personal bar soap market (at the time the research was conducted). Similarly, we estimated touch to account for $34 million and sight $14 million. On a combined basis, users' memories of the sensory experience of Dove account for around 25 percent of Dove's annual revenues in the United States—and that is just for the United States alone.

These estimates are crude. They do, however, start to give some sense of the scale involved. Although as marketers we focus much of our attention on communications and positioning, we cannot afford to dissociate them from the experience of the product itself; much of a brand's value rests in consumers' memories of their experience with it. Imagine you were considering changing the Dove formulation to a cheaper recipe that would save $5 million a year but would negatively affect the user experience. Might the suggestion that the decision would affect over $100 million in revenues based on the sensory experience give you pause?

USING INNOVATIVE DELIVERY TO GOOD ADVANTAGE

I first learned about Blue Moon, a Chinese brand of laundry detergent, on a recent trip to China. I had seen the brand name in our BrandZ database but had not really known much about it. Blue Moon turns out to be characteristic of a new crop of Chinese brands that are willing to take on the multinationals and charge a premium price for their product and are not averse to adopting aggressive marketing tactics to do so.

Launched in 2008, Blue Moon is a liquid detergent targeted to the segment of consumers who prefer to wash clothes by hand. Multinational brands like Unilever's Omo, the long-standing detergent category brand leader, have typically focused on laundry powder. In contrast to other Chinese brands' images, Blue Moon is not perceived as cheap; it is seen as different and desirable. Moreover, Blue Moon is not afraid to invest in marketing. In 2011, Blue Moon doubled its ad spending to more than $500 million, grabbing a 30 percent SOV that is far in excess of its overall category market share. As a result, the brand's market share has grown rapidly over the last five years.

When Blue Moon suddenly stopped spending on above-the-line media early in 2012, marketers responsible for the major multinational brands breathed a sigh of relief. That's it, they thought, Blue Moon has run out of money, it is on its way out. In China's frenetic marketplace, where brands need to ensure they remain salient, the lack of advertising seemed like a death sentence for the brand. However, nine months later, Blue Moon was still going strong, with no sign of eroding attitudinal equity or sales.

So what was going on? It turned out that Blue Moon had taken a radical step indeed. The money that had been invested in above-the-line media had been redeployed to in-store activation. But this was not the typical in-store display and price promotion. Instead, Blue Moon was offering consumers a completely different and meaningful brand experience. While other brands were investing in TV advertising and reaching out to consumers online, Blue Moon promotional girls were busy engaging people at the point of purchase and encouraging them to smell the brand as if it were a fine perfume. Asked about the experience by Millward Brown researchers, one shopper reported: "The girl does not only tell me [Blue Moon] has good fragrance, but also teaches me how to smell it properly—not directly from the bottle but wafting the hands back and forth in the air to enjoy the fragrance."

What a fantastic way of engaging people with a brand to which they might not have paid attention otherwise. In China, people are eager to learn more about brands and are used to seeing promotional girls in-store, but Blue Moon's approach sidesteps the usual explanation of functional benefits and price promotion to get people to interact with the brand directly.

Our sense of smell has an incredibly strong, sometimes unperceived influence on how we react to the world around us and is particularly good at eliciting a positive emotional response. It is tough to tell whether your clothes are as clean as they could be, but you can certainly tell whether or not you like a scent. By disrupting the rules of the category, Blue Moon provided potential consumers not just with a sales pitch, but with a new learning experience.

The big question, of course, is whether this approach will be effective over the long term. Competitors are bound to respond in some way. While retailers may balk at having multiple brands competing for peoples' attention in-store, competitors will do their best to blunt Blue Moon's advantage. It seems likely that having created the disruption, Blue Moon may choose to return to above-the-line investment until they can figure out a new way to sidestep their competition.

THE CHALLENGE OF DELIVERING GREAT SERVICE

In 2008, I heard Chris Clark, head of marketing, planning, and business strategy at HSBC, give a presentation at the inaugural International Advertising Association Global Brand Think Tank held in the United Kingdom. Clark shared an amusing anecdote to highlight the challenge of marketing a bank, recalling the time that a human resources manager called him and complained that HSBC's advertising was creating the wrong expectations among new recruits to the bank's U.K. branches. The manager told him that the recruits thought they were joining a dynamic, cutting-edge company, but after a year, many left, disillusioned by their actual experience. Clark gave the guy short shrift, saying, "Well, we have two choices. Either you make the bank a better place to work, or I can create a worse brand."

Therein lies the fundamental problem of any service brand. Expectations created by marketing must be met if customers and staff are to be satisfied. As Clark put it in his presentation—referencing the warning commonly heard in the London underground—service brand marketers need to "mind the gap."[5]

That gap was thrown into sharp relief in the United States recently when the insurance and financial services multinational AIG managed to send very mixed messages to a public that already doubted its trustworthiness. On one hand, the company launched a new advertising campaign, "Thank You America," which sought to show how the brand had paid back its bailout money, cleaned up its act, and reached out to help people suffering from the aftereffects of natural disasters. On the other hand, AIG announced that its board was considering joining a lawsuit against the federal government, claiming that shareholders didn't receive fair compensation when the company was bailed out. The negative publicity resulting from that announcement undermined the intent of the advertising campaign, even though AIG did not ultimately join the lawsuit.

An expectations gap can exist for any product or service, but it is more common for people to be dissatisfied with a service brand. Consider the following statistics pulled from a sixteen-country analysis of BrandZ data:

- Because of the risks associated with the choice, people are more likely to believe it is important to choose the right service brand than the right product brand. On average, 57 percent of people say it's "very important" to choose the right brand of a service, compared to 42 percent for product brands.
- People are less likely to be satisfied with their choice of service brands than their choice of product brands. On average, only 26 percent of people are totally satisfied with their service brands, versus 40 percent for product brands.
- Fewer people are attitudinally loyal to service brands. On average, 46 percent of category users are bonded to a service brand (i.e., to any brand in the category), compared to 57 percent for product brands.

So why do these differences exist? I think there are three main reasons why service brands tend to be weaker than product brands.

1. Service brands rely on systems and processes that are often poorly aligned to serve people's needs, particularly when things go wrong

Take my experience of a credit card being cancelled because of possible fraud. A recent merger had left the systems of my financial company in disarray. When I attempted to get a new card issued, I entered call-center hell, as I was passed from one unit to another, trying to find the right person to help me. A similar but less important problem occurred when United merged with Continental and my frequent flier records were confused.

2. Human interaction is intrinsically variable, and person-to-person exchanges with a company employee can create strong emotions—good and bad

Last year, Gordon Pincott checked into a five-star hotel in Colombo, the capital of Sri Lanka. It was a frustrating experience of waiting, misinformation, and needing to change rooms. As he was finally settling in, the guest relations manager called to check whether everything was all right. This question elicited a detailed response of the failings of the hotel. What became clear was that the guest relations manager had a job description that required giving every new arrival a courtesy call half an hour after he or she had checked into a room. It was not the manager's role to listen to criticism, and she was not empowered to do anything about it. So her response was to get off the phone as quickly as possible and move on to the next call. An audit would have revealed that a guest relations process existed, but it would not have revealed that it failed to work at the times when it really mattered.

3. When things go wrong, service brand customers often have no recourse to change vendors and often feel undervalued, trapped, and resentful

Service brand customers often feel that they cannot easily switch to a new provider because of contractual arrangements, the work involved in making the change, or the potential fees they may incur. On average, 13 percent of people report that they are not satisfied with service brands, compared to only 4 percent for product brands. These people do not necessarily behave any differently in terms of their use of the service, but they may well talk about it differently: service brand customers are 38 percent more likely to recommend against a brand than are product brand users.

Given these findings, it is not surprising that service brands tend to be weaker than product brands. A strong service brand must have robust business systems that are aligned to meet customers' needs and a staff that is equipped with the knowledge and the authority to handle issues as they arise. Neither one of these tasks can be accomplished overnight, but if they are done right, then the human element—the factor that can make customer experience of a service brand so variable—can be used to great advantage. By effectively handling the human elements of the business, a service brand can set itself apart as something truly special.

Consider the difference between buying a package of butter and making an insurance claim. The act of buying the butter is unlikely to be disappointing; it's a predictable, low-stress transaction, likely to be completed satisfactorily almost every time. But that package of butter has relatively little power to alter the mood of the buyer, to make him or her smile or laugh, or to impact his or her day in a new, meaningful way.

In contrast, a rewarding human interaction, particularly at a stressful time, can leave a significant impression on a customer. In the United States, the State Farm insurance company uses their advertising to highlight the power of human contact in stressful times by featuring the testimonies of satisfied customers, all of whom finish their stories with the company's tagline: "Like a good neighbor, State Farm is there."

One last point: many service brands religiously track their customer satisfaction and seek to address problems when they are identified. However, there is a big difference between customer satisfaction and customer loyalty. A customer may be satisfied with a series of successful transactions but not feel loyal to the brand. Marketing is not just about creating expectations; it is also about creating a willingness to forgive and forget when things do go wrong. Good marketing communications can help insulate the brand by reminding customers of the good experiences they've had with the brand and making sure that they understand what the company is trying to do right, even if it does not achieve it all the time.

IS DESIGN THE NEW MARKETING?

In 2010, Procter & Gamble incorporated design into its global brand-building function under Global Design Officer Phil Duncan. The objective of this change was to ensure that product, packaging, and marketing were aligned to deliver a great brand experience. Part of this change was a recognition that the most meaningful brands in the world are built not on advertising, but on personal experience. In a lecture at Northwestern University's Robert R. McCormick School of Engineering and Applied Science, Duncan put it this way: "We really are trying to understand how we can build a one-on-one, meaningful relationship with every person in the world. Pretty daunting task, but again, technology and insights enable us to do that."[6]

How better to create a one-to-one, meaningful relationship than through the actual brand experience? As Duncan stated in an earlier interview with

AdAge: "We're paying attention, finally, to the things that matter to consumers, and stripping out the things that don't."[7]

These words also reflect the philosophy behind Apple's most successful products: the iMac, iPod, iPhone, and iPad. Talking about his design philosophy, Jonathan Ive, senior vice president of industrial design at Apple, says of the Apple products, "At some level, I think you are aware of a calm and considered solution that therefore speaks about how you are going to use it, not the terrible struggles that we as designers and engineers had trying to solve some of the problems.... That's quite obsessive isn't it?"[8]

The funny thing is that when many marketers think about the success of the Apple brand, they often focus on its looks, cool advertising, and perceived popularity—not on the brand's inherent usability. They focus on the effect, not the cause. The real cause of Apple's success is the natural, intuitive transparency of the technology. The superlative product experience comes from an unusual combination of human and technical understanding, and this creates the foundation for all the other positive aspects of the brand.

Many companies would do well to learn from Apple's example, including Ford. At a 2012 Los Angeles auto show, Ford's president of the Americas, Mark Fields, claimed that his company was thinking more like a technology company, but the recent response to the MyFord Touch suggests that they still have a ways to go achieve the same competency with technology that they have achieved with cars.[9]

The MyFord Touch system—introduced in 2011 as a way to stand out from the rest of the industry—actually led to a decline in Ford's initial quality ratings as measured by J. D. Power and Associates. In contrast to the iPod Touch, which is a great steak with lots of sizzle, Ford's MyTouch seems to be a lot more sizzle than steak. While the idea of MyFord Touch was initially appealing to buyers, more long-term experience of using the touch screen system was far less positive. Since its introduction, Ford has upgraded the system and extended its warranty from three to five years, with unlimited mileage. This move is less a sign of confidence in the reliability of the system and more a safety valve that allows owners to upgrade their system as its problems are addressed.

Design is the practice of creating a positive and enjoyable experience that engages all the senses and adds meaning to people's lives. This is not something that applies only to product brands. Any brand can benefit from a design process that focuses on creating a simple and effortless experience. Today, people are used to products and services that meet their needs *well*. The real challenge is to create a brand experience that is so good, so satisfying, that it is almost invisible. Design is about creating a desirable experience, one that users want to repeat, and if loyalty originates from good design, then marketing must embrace design as a core competency if it is to be effective.

■■■■ EFFECTIVE DELIVERY IS CRITICAL

The importance of good quality and service is well recognized; obviously, no one is going to buy a brand they find unsatisfactory when suitable alternatives exist. But all too often, a company's definition of their "product" is far more limited than that experienced by the consumer. A bad packaging experience can detract from the enjoyment of a food product. A bad customer service experience can set a customer down the road to finding a new provider. Frustration with a balky user interface can detract from an otherwise perfectly acceptable product experience. Pain points like these need to be identified and eliminated.

Perhaps even more important is the added value created by harmonizing a brand's touch points with what makes the brand meaningfully different. Communicating what the brand stands for is important, but when people experience that difference by interacting with the brand, the understanding becomes visceral. The alignment does not need to be explicit in every instance, but it should at least be implicit in design, sensory experience, and tonality.

Questions to Consider

1. Does your brand deliver on its purpose? Is the brand experience a good one? Review up-to-date product performance data against that of the current competition. If you are a product brand, this review should take the form of consumer product testing. For service businesses, you might use customer satisfaction studies or mystery shopping. For online businesses, review could take the form of accompanied or monitored browsing. Make an objective evaluation of the quality of your delivery. When possible, you may want to adopt a de-branded, as well as a branded, approach to allow you to disentangle the component parts of the delivery—to separate the tangible from the intangible.
2. What needs to be done in order to improve the way that the product or service is delivered? You will need to understand which improvements will realize the biggest benefits to your consumers and the best financial return. This information will inform and prioritize innovation, product development, and remedial actions.

Chapter 7

Resonance

A clearly defined purpose—and a company organized to deliver that purpose—will do little to build brand value if the brand's offerings are not useful and relevant to consumers. The brand must offer something that consumers need or want at a price they are willing to pay. But strong brands are built on more than creating satisfied customers; strong brands also strike an emotional chord with consumers. Strong brands resonate.

The most powerful type of resonance occurs when strong emotional bonds connect consumers with brands. These emotional connections are formed when people feel that brands are not only useful, but also emotionally rewarding. This is the point at which the intrinsic and extrinsic elements of the delivery combine to make the brand significant to people.

Abraham Maslow was an American psychologist made famous by his description of human needs as a pyramid rising from a base of physiological needs up through safety, love and belonging, and esteem to the pinnacle of self-actualization. Brands build resonance through a hierarchy of meaning that roughly parallels Maslow's hierarchy of needs: functional, emotional, social, and cultural. Unless a brand can ensure that it is conveying the lower levels of meaning, it will fail to achieve the higher levels. You can argue about whether or not Patagonia's outdoor gear is really different from the gear sold by North Face or Arc'Teryx, but the social values of the company definitely are. Would Patagonia be successful if it failed to deliver great quality products? No. The company's ethos would still strike a chord with people, but most would not buy the products.

Consumer resonance becomes more difficult to achieve as the definition of the target audience broadens, and the goal of marketing is to make the brand as relevant as possible to as many people as possible without losing clarity about what the brand stands for.

ADDRESSING A NEED

In 2001, American Express Small Business Services faced a crowded and undifferentiated marketplace. Even though American Express had pioneered the small business charge and credit card market, banks had managed to leverage their existing business relationships with customers to muscle in on the company's territory. In spite of providing a wide range of business services, most small business owners associated American Express with credit cards and, worse, tended to limit their use of business credit cards to travel and entertainment. Banks seemed a more logical source of financing and other financial services, even if the experience of dealing with them left many business owners feeling dissatisfied. The question facing American Express was how best to capitalize on its credit card business without being defined by it. The OPEN Small Business Network was the solution, and its success was the result of a compelling insight into the lives of the target audience.

Jeremy Diamond, then a senior partner and worldwide group planning director at Ogilvy and the man responsible for brand strategy for American Express, explains the motivational power of the OPEN idea this way:

> The key to OPEN was the insight. Small business owners set up their businesses to take control of their future, but as a small business owner they often felt insecure and vulnerable. This insight into the emotional need drove everything. Most importantly, it drove the brand promise, that American Express would give them a feeling of control over their future. This became the lens through which we defined the whole of the OPEN brand experience. It drove the product proposition—American Express would provide a complete network of resources to allow small business owners to keep control. The cards became keys to this network. It also drove the service experience—by asking when this emotional need was most pronounced, we could identify the moments of truth when small business owners were most in need of, and most responsive to, the support the brand could offer. By giving them back a feeling of control at each of these moments of truth, American Express became more than just another payment card; it became an indispensible ally that they knew they could depend on.[1]

"Open" was originally the creative idea designed to highlight the divide between the attitude of American Express (open) and the attitude of banks (closed), but it quickly became apparent that the power of the word itself could be more than descriptive. OPEN also become the name of the subbrand that would set Small Business Services apart from the card-only positioning of its parent company. Diamond states, "What we learned from

OPEN is that the single most important question you can ask yourself is, 'What is the underlying emotional need that my brand can best address?' Once you have this, you know what your brand can mean to people. This drives every aspect of the brand experience."[2]

OPEN became the lens through which every aspect of the brand experience was defined. American Express and its agency partners set about redesigning their website to make it easier for small business owners to find solutions that met their specific needs and allowed them to network with their peers. Charge and credit cards were redesigned and rebranded as "accounts" to reflect the fact that they were no longer just for payments, but would also act as membership cards giving access to a range of financial resources. Within American Express itself, call center staff were retrained, and scripts were changed to reflect the new ethos. A 360-degree campaign was designed to highlight the OPEN Network as an advocate for the small business community, with subsequent executions focusing on how the OPEN Network empowered small business owners at critical "Moments of Truth." The OPEN Network launched in January 2002 and immediately gained traction.

■■■■ BUILDING ON STRONG FOUNDATIONS

Provided that the idea is well executed, an insight like the one that led to the creation of the OPEN Network is the gift that keeps on giving. Because the needs being addressed are timeless, a brand can evolve with its customers over time without changing its basic positioning.

In 2010, American Express OPEN introduced Small Business Saturday as a way to tap into the widespread but disparate support for local businesses. Small Business Saturday was designed to offer a counterpoint to Black Friday (the day after the Thanksgiving holiday when big-box retailers begin to turn a profit for the year) by focusing attention away from the big-box stores and toward the small business community.

American Express helped launch the day in 2010 by providing free Facebook ads to merchants and a $25 rebate to American Express customers who shopped at small businesses. Not surprisingly, given the economic times, the campaign resulted in 1.2 million people liking the Small Business Saturday Facebook page in 2010. With further efforts in 2011, that number grew to 2.7 million, and support spread beyond small business owners and consumers to big businesses and politicians, including President Obama, who tweeted in 2012: "My family & I started our holiday shopping at a local bookstore on #SmallBizSat. I hope you'll join & shop small this holiday season."

According to a survey by American Express and the National Federation of Independent Business (NFIB), in 2012, consumers spent an estimated $5.5 billion at locally owned stores and restaurants on Small Business Saturday, no doubt giving both small business owners and American Express a welcome boost.[3]

ADMIT FAILURE AND GAIN MOMENTUM

Admitting you are wrong takes courage, particularly when there are millions of dollars in revenue on the line. But if the alternative is watching sales and share prices erode, then admitting you are wrong may be the right thing to do. This was certainly the case for Domino's Pizza.

In December 2009, Domino's launched an ad campaign in the United States that highlighted just what people thought of the company's pizza: they thought it sucked. The ads acknowledged this long-standing problem, announced that the company was trying to change its product, and asked for a second chance. This radical approach reaped significant rewards. Following four years of flat or declining sales, same-store sales grew 14.3 percent in the first quarter of 2010 compared to the same period the previous year, significantly outpacing the competition. Overall, in 2010, U.S. same-store sales rose 9.9 percent in a mature market, in which 1 to 3 percent growth would be the norm.[4] Results from the third quarter of 2011 found same-store sales in the United States up 3.3 percent, indicating that the brand had held its gains from the previous year and that sales momentum was continuing.[5]

Central to the success of the new ad campaign was the recognition that admission of fault would resonate strongly with people familiar with the Domino's brand. Set in the context of the Great Recession, qualitative research found that people displayed an instinctive distrust of big business and institutions. People complained that companies never admitted they were wrong; they never simply said, "Hey—we screwed up." The Ogilvy Awards case study notes: "Admission is interesting. It's humanizing. When a company admits they're wrong they begin to seem human, fallible and vulnerable.... But most of all, admission lays the foundation for a new relationship. It's like a reset button."[6]

Scripts admitting the company's past failures and announcing plans to change for the better were taken into creative development focus groups. There, the idea of a company admitting fault proved to be incredibly powerful, giving people a reason to suspend judgment and reconsider their understanding of the Domino's brand. Further research refined the campaign and focused on the simple idea that instead of denying criticism, Domino's listened, responded, and created a better pizza. To enhance the authenticity of the campaign, it featured Domino's CEO J. Patrick Doyle and other executives telling the story of how they responded to criticism and developed a better pizza.

Even with all the evidence in support of the strategy, acting on it still required considerable resolve on behalf of Domino's management. As Doyle acknowledged in an interview with Bloomberg, the announcement effectively burned the bridge behind them—once it was spoken, there was no going back.[7] As he admitted in an interview with *Time* magazine, "I'd be lying if I told you I didn't have a knot in my stomach the day that advertising launched."[8]

Nevertheless, the decision proved worthwhile, because the campaign resonated with consumers. With sales growing once again, Domino's share price, which had hit a record low of $2.83 a share in November 2008, rebounded to a four-year high of $28.87 per share in September 2011.

Perhaps one of the reasons that the Domino's relaunch worked so well is because it did not try to hype or spin the reality of the situation; instead, Domino's admitted their failures and explained how they had improved their product. The fact that the company could present itself as having done something concrete to respond to customer discontent stands in stark contrast to many brands that put claims before action.

GREAT PRODUCT, LOUSY MARKETING?

Would you recognize a Millennial if you saw one? Millennials, who used to be called Generation Y, were born between 1982 and 2004. Compared to previous generations, Millennials are supposedly more self-centered; more focused on becoming wealthy; and more familiar with communications, media, and digital technologies. Faced with declining canned soup consumption and declining market share, Campbell's, best known for its iconic red and white label, created a new line called Campbell's Go™ Soups that were designed to appeal to Millennials.

Campbell's problem is familiar to any long-standing brand: the people who buy their soup are aging, and new generations no longer find the brand as relevant as it once was. Today, people's eat-on-the-go lifestyle is well served by a vast range of fast food outlets. The types of food that people enjoy represent a far more global palate than they used to, as "ethnic" flavors from Mexico, the Middle East, India, and China have become popular. When people eat a quick meal at home, the microwave is likely to take a front and center position in food preparation, not the stovetop. Moreover, memories of Campbell's canned soup are not necessarily positive. In a world that increasingly values fresh and organic foods the cans themselves have become synonymous with "stale" and "artificial." Finally, the variety of packaged soups on the market is far greater today than it was back in the 1960s. Campbell's alone offers many other lines in addition to its original Condensed Soup line, and there are many other brands besides, several of which offer alternative packaging, like Tetra Paks.

If Campbell's is to be criticized for its decline in market share, then it is because they did not address these trends sooner. However, having identified the need to do so, they pulled out all the stops. An article by Candice Choi in the *Huffington Post* describes the extensive consumer research conducted in order to understand Millennials and how best to appeal to their tastes and lifestyles.[9] Much of the company's research appears to have consisted of ethnographic research, spending time with the target group at home and in restaurants. As Choi notes, "The fieldwork led executives to two seemingly divergent conclusions: First, cuisines once considered exotic—Thai, Indian, Brazilian—have become the norm. At the same time, years of dining out mean younger consumers aren't as skilled at making meals from scratch, particularly when it comes to those very ethnic flavors."[10]

Whenever there is a tension between behaviors like this, there is also an opportunity. Campbell's Go Soups seek to address that opportunity by offering more exciting soup recipes in easily microwaveable pouches. I don't know about the Millennials, but with varieties like Creamy Red Pepper with Smoked Gouda, Spicy Chorizo & Pulled Chicken with Black Beans, and Coconut Curry with Chicken & Shiitake Mushrooms, the new line sounds good to me (a trailing-edge Boomer). What is less appealing, however, is the way that Campbell's has gone about marketing these soups. Their approach is a classic example of pandering to the superficial behaviors of your target audience without connecting that marketing to the actual product experience.

Well-known marketing expert—okay, comedian and wise guy—Stephen Colbert also knows a crock campaign when he sees it. On his satirical news show, *The Colbert Report*, Colbert took the time to examine the Go Soup strategy and campaign. Explaining that he's a part of the TV biz, which is all about the demographics, he states, "So I work hard to appeal to the Millennials—for example, by calling them Millennials. Young people love being targeted by their birth date and purchasing power."[11] Colbert then proceeds to parody his attempt to appeal to Millennials by throwing contemporary slang and gestures into the remaining minutes of the sketch. Here is what Colbert has to say about the Go Soups marketing campaign:

> And because it is so hip, homeslice, you're not going to see ads on the TV, no that's for squares, you've got to surf over to the Campbell's website, with edgy Tumblr-inspired graphics that capture all the rebellious spirit and relentless energy one associates with soup. Of course, you're probably thinking, what's soup without music? That's what I thought. Well they've got that covered too. 'Cos Campbell's has partnered with the music service Spotify so that [quoting from a *Mediapost* article] "consumers can create custom playlists built off the persona of the soups."[12]

What Colbert clearly recognizes in his report is that the Go Soup marketing campaign is designed to resonate with Millennials but totally fails to connect their interests with the product being sold. Colbert concludes his report by referencing the brand's playlist partnership, saying, "For instance, I chose Tom Petty's 'A Wasted Life' to describe whoever came up with this marketing campaign."[13]

That might sound harsh, but I have to agree that the campaign seems to owe a lot to the latest hot trends in marketing and precious little to the goal of selling soup. The superficial appeal to Millennials' interest in social media and music is typical of many marketing campaigns these days. While it might create some trivial buzz, the campaign fails to answer the fundamental question of why anyone would want to buy the soup, apart from the lure of free stuff and a $1-off coupon. Rather than making the product attractive, it seeks to incentivize people to engage with the brand, and even then the campaign fails to resonate with its target audience.

Think this is just the viewpoint of an aging pundit? Let's hear what Maggie Hayn, a Millennial and member of Millward Brown's U.S. client service team, has to say:

> My theory is that Millennials are more aware of the difference between a wishy-washy campaign and a tightly put together campaign, because we develop campaigns for our personal brands every day. We post images and clever sayings and opinions for the world to see, and we're careful about the effect of each of these on our overall public/online image. We take into account how it will be interpreted by family, friends, and employers. A Facebook post has to be interesting enough to grab attention, witty enough to make you sound smart and funny, and politically correct enough not to offend anyone. It has to tie in to what you stand for as a person, and it can't contradict anything you've said before. The best Facebook posts generate positive conversation with spirited commentary beneath it. These are very similar ideas to how a

brand would want a campaign to work. When we [Millennials] work so hard on projecting our own brand image and then see a brand like Campbell's, which has teams of experts with big budgets, simply failing at projecting their intended brand image, it makes them appear weak and clueless and turns us off to the brand.[14]

I think Maggie has identified why so many brands fail to resonate with their customers and consumers today. They know their target audience's behaviors, but they fail to go beyond those behaviors to understand real motivations. They lack the ability to see the world from someone else's shoes.

Will Go Soups prove successful? Only time will tell. My concern is that although the soup recipes may well appeal to contemporary tastes, other fundamental shifts in consumer behavior remain unaddressed. Can people be bothered to go home and heat up a pouch of soup when they can pick up a quick, tasty burrito or soup on the way home? It seems to me that Campbell's needs to find ways of positioning the brand as relevant to Millennials in terms of motivations and lifestyle at a more fundamental level—perhaps by positioning Go Soups to appeal to cash-strapped Millennials as a cheaper alternative to takeout? Or maybe Campbell's needs to get ahead of the next culinary curve by offering fewer meat-based soups and more vegetable-based ones instead. But then again, that strategy might take a bit more investment in the brand than Campbell's seems willing to make. In seeking to engage Millennials with the brand, the campaign totally ignores the need to reach as wide an audience as possible and let them know that Go Soups exists.

EMPATHY AND AUTHENTICITY

I think the keys to creating a brand that resonates are empathy and authenticity. These are the qualities that will help differentiate a superficial brand from an emotionally resonant one. Empathy with the target audience is the starting point. A deep understanding of the passions, concerns, and emotional needs of the people your brand serves will help identify what makes this brand truly meaningful to them. In the case of American Express OPEN Network, the key was to recognize the underlying anxiety suffered by small business owners and to find a way that American Express could address that need. Authenticity, staying true to the brand's purpose and values, will ensure that people respond well to how the brand presents itself. Domino's was successful because it faced up to a problem and addressed it publicly, rather than trying to gloss over it. The honesty of this approach resonated with the public and made people willing to give the brand a second chance.

Questions to Consider

1. Do people appreciate what your brand offers? To really understand what your brand could mean to its consumers, you need to be able to see it from their viewpoint. Personal observation, ethnographic research, and other qualitative techniques can all give you an understanding of people's explicit cognitive reactions toward a brand. Neuroscience approaches can uncover the implicit associations and emotional connections that people have with the brand. This information will be particularly helpful in situations in which people have reasons to hold back on expressing their explicit responses. These reasons could be cultural or might have to do with the sensitivity of the product category, or perhaps it may simply be hard to articulate exactly what someone feels about a brand. All of these data would need to be put into the context of a competitive set in order to compare results against benchmarks or databases or competitive brands covered in the same study.

2. Does your brand resonate with its target audience? Does that audience appreciate what the product does for them? Do people feel emotionally rewarded by the experience? Better yet, are they inspired by it? If so, what makes them want to recommend it? What would help the brand resonate better? What would inspire more people? Explore ideas and avenues to improve how well your brand resonates with its target audience using in-depth qualitative research techniques.

Chapter 8

Differentiation

"Differentiation is the essence of strategy."

So write Chris Zook and James Allen in a recent *Harvard Business Review* article.[1] Joan Magretta puts it this way on the *Harvard Business Review* blog: "Nothing is more absurd—and yet more widespread—than the belief that somehow you can do exactly what everyone else is doing and yet end up with superior results."[2]

Zook and Allen identify fifteen basic categories that underpin a differentiated strategy. These fifteen are grouped into three macro-level categories: management systems, operational capabilities, and proprietary assets. Brand is included as a proprietary asset, but, as I have noted elsewhere, a brand is a two-sided coin. It is not detached from the business; it is a manifestation of it. What your customers or consumers experience will inevitably reflect management priorities, production capabilities, supply chain logistics, research and development, customer service, and marketing communications. Therefore, if what is said and shown in your marketing diverges from the overall strategy, then there is likely to be a significant problem. Either the brand overpromises—as in the HSBC case cited in Chapter 6—or it creates a superficial allure that is detached from the brand experience. Neither strategy is productive in the long term.

███ USING YOUR BRAND TO CREATE A SUSTAINABLE ADVANTAGE

If a brand is intended to create a sustainable competitive advantage, then that brand had better be well differentiated, and marketing must reflect and enhance that differentiation. Think about the brands that are typically cited by marketers as the ones they admire: Apple, Virgin, Facebook, Disney, Coke, Audi, Jack Daniel's. These brands are all well differentiated; they set the trends for their category or transcend it, and they act differently and stand out from their competition. Most of these brands are differentiated on both a tangible and an intangible basis. Whether the differentiation is functional, physical, or visual, or whether it is based in persona, tonality, or viewpoint, these brands are not going to be mistaken for another. In relatively undifferentiated categories, marketing—how the brand communicates—becomes the primary source of differentiation. We will consider cases like those in more detail later in this chapter.

If a brand is going to be a strategic asset, then the differentiation created by the brand needs to be sustained, meaningful, and easily appreciated. Ideally, that differentiation will be founded in a tangible difference. Why? Because it is far easier for people to confirm what they have been told if they can see, feel, taste, smell, or hear the difference. That difference may not be very big or even important, but by drawing attention to it, good marketing can magnify its influence and make it important. And if the difference is intangible, then it better have the potential to resonate with the target audience. They need to feel it emotionally. If your consumers don't feel it and believe it, then you are not different, no matter how different you think your brand might be.

We live in a complicated, noisy world. Brands compete for attention and affection with the myriad other things going on in people's lives. There is not much chance that people will remember a lot about our brands—as we learned earlier, people tend to respond to brands based on very simple associations—so we need to make sure that people know what makes our brand different from the competition.

THE HIDDEN POWER OF DIFFERENCE

In Chapter 2, I described how brands influence purchase decisions, and we learned that difference is a necessary criterion for choice. A meaningful difference—no matter how trivial—provides a reason for people to stop assessing alternatives and choose one of them. After that decision has been made, the difference itself may become "unimportant." In other words, once a brand has been chosen, it becomes far more important that the brand do its job and that people like it. The reason for their choice may no longer matter to people. They have made their decision and are going to stick with the chosen brand unless something else happens—their needs change, they have a bad experience, or they become aware of a better competitive offer.

If you conduct what market researchers like to call a *"driver analysis,"* you will see that some attributes or associations always seem strongly related to why people choose a brand. Rather than ask people to rank attributes by importance, a lengthy and potentially misleading approach, driver analysis simply correlates the attributes with an outcome like "brand bought most often." The correlations identify which attributes are more likely to be associated with purchase. Unfortunately, while easy to apply, this approach also suffers from drawbacks, not the least of which is that the analysis is not really telling us much about the real motivations for purchase.

For instance, in food categories, associations like "good taste" and "most popular" are usually strongly correlated with choice, and others like "different" and "growing more popular" rarely show up as important. The problem is, however, that while it is important that food taste good (unless there is an overriding other reason for eating it, like heart health), good taste is a basic requirement of any food brand. It does not offer competitive advantage, only the avoidance of disadvantage. "Most popular" is simply a descriptive statement. (Popularity does confer an advantage, but it is not one readily recognized as "important" by respondents.) Big brands, the ones most people buy, are always the most popular.

How do brands become the most popular? They do so by growing in popularity. People can recognize when a brand has momentum and is growing more popular. That perception is unrelated to their current brand choice. However, the perception that a brand is growing in popularity may cause people to pay more attention to that brand, and, when the occasion arises, they may check out why it is becoming more popular. Later, we will consider the example of Audi in the United States, where the creation of the perception that the brand was growing in popularity boosted sales still further.

If someone uses a food brand, then he will usually mention it in relation to all the basic category requirements—good taste, convenience, appeals to the family—because he would not use it otherwise. That person may not think to mention that it is different from other brands because he does not evaluate it against the alternatives every time he shops (the exception being price shoppers, who will check for discounts and deals). If someone does not use the brand, then either she was unaware of a meaningful difference that made her want to try the brand, or, when she did check it out, she found it to be different in a bad way. Only by talking to people immediately after they have made a decision can we really hope to identify the true scope of difference in people's decision-making; otherwise, we are limited to those cases in which the brand has created sustainable and widely understood differentiation.

■■■ THE OBVIOUS POWER OF DIFFERENCE

Salience is the property by which brands come to mind quickly and easily. Our research confirms that salience is best measured in association with category need-states. The idea of a *need-state* recognizes that the need for a brand extends beyond the physical. Depending on the circumstances, your choice of beer may be affected by factors that include the occasion or psychological need. The light, refreshing beer that seemed so enjoyable when chugged down after a hard day's labor might seem insipid when relaxing by a winter fire. When socializing with friends, you might choose to drink the same brand as they do in order to fit in, or you might pick a different brand in order to stand out.

Take the example of airlines in the United Kingdom, where British Airways (BA) is a very salient brand. In research conducted to pilot our new approach to brand equity measurement, BA was the strongest brand in terms of traditional top-of-mind awareness, or recognition triggered by the category name. However, when we applied a needs-based approach to salience, easyJet emerged as the most salient brand. That's because easyJet has built a strong association with low-priced air travel, a widespread and strongly felt need among vacation fliers. The airline differentiates itself on the basis of price and has amplified that difference effectively.

Red Bull is an amazing brand that has done a fantastic job of ensuring that it is associated with a specific need-state. The brand seeks out situations in which people are physically and/or emotionally drained in order to introduce them to the product. Whether it is after climbing a tower in the old city of Bruges or arriving jet-lagged in a different country, a shot of Red Bull delivers enough sugar and caffeine to give you a boost—even without the taurine, glucuronolactone, and B vitamins also included in the drink. Most people do feel revitalized after drinking Red Bull, and the brand's unique approach to high-adrenalin sports helps ensure that people who want to be mentally and physically active remember that Red Bull can give them wings.

The result of Red Bull's well-thought-out and well-implemented marketing strategy is that the brand is salient in relation to specific need-states. If people want refreshment, they will turn to Coca-Cola or another brand, but if they want to be energized, they are more likely to turn to Red Bull and be willing to pay a premium for it. *Euromonitor* reports that Red Bull sells at a 50 percent premium over other energy drinks, a category that, on average, maintains a price point per liter nearly five times that of regular carbonated soft drinks.[3]

DIFFERENCE CREATES POTENTIAL FOR GROWTH

The importance of difference in decision-making is also apparent when we look at its role in determining whether a brand has the potential to grow market share over time. The biggest driver of growth is how well a brand can make new consumers aware of what it has to offer, and how well a brand capitalizes on that increased awareness depends on how "sticky" the brand's offer proves to be. In our research, brands that people find to be meaningful and different are more likely to be chosen and are more likely to satisfy people's needs over the long term than others. We find that brands that have already proven their ability to convert people from familiarity with the brand promise to perceiving the brand as meaningful and different are the ones most likely to grow market share in the year following the survey.

DIFFERENCE JUSTIFIES A PRICE PREMIUM

Being perceived as different will help people choose your brand when faced with similar alternatives, but the more obvious and sustained benefit of that perception is that it helps justify a price premium. Perceived differentiation allows people to say, "It's worth paying more for this brand because it gives me something I cannot get elsewhere."

For an example of great brand differentiation, we need look no further than the most valuable brand in the world, Apple. It should come as no surprise that across the categories in which it competes, Apple products consistently over-index as meaningful, different, and salient. However, it is the brand's differentiation that is outstanding. Perceptions of Apple's difference are founded on its product innovation and delivery, but Apple goes beyond product differentiation to project a unique personality and set of values (BrandZ finds that Apple consistently over-indexes on having a different personality). It is also the strongest global brand in the database for perceptions of having company ideals.

The power of difference to help justify a price premium does not preclude a value brand from benefiting from being seen as different. The furniture retailer IKEA is a case in point. In the United Kingdom, for instance, the brand is seen as different, trendsetting, and a better value than other furniture stores. IKEA does not seek to appeal to everyone, but it does aim to satisfy those who need to furnish an apartment or house as quickly and cheaply as possible. In China, the e-commerce site Taobao is used by millions and is perceived as different and as offering better prices than its competition; in India, the same applies to Colgate, which effectively markets its brand across a wide price range. Our research suggests that these brands could command a higher price point than they do, but they are willing to forgo premium pricing in favor of driving volume sales.

■■■■■ SETTING THE STANDARDS FOR COMPETITION

"Physical" differentiation is the domain of good design. This area includes product and service functionality plus the aesthetics of product and packaging. It has become popular to state that functional difference is impossible to sustain in today's hypercompetitive marketplace, but, as the example of Apple demonstrates, it is possible to create sustainable differentiation.

The iPod significantly upstaged the Walkman and redefined the standards for portable music so completely that the Walkman became essentially redundant. (Although production of the Cassette and MiniDisc Walkman has been discontinued, the CD Walkman remains in production a decade after the introduction of the iPod. It takes time for evolution to run its full course.) Apple was not the first company to introduce a digital music player, and many companies have since introduced their own, but none of them compare to the iPod.

There are three reasons why Apple has been able to maintain its preeminent place in the category.

1. Archetypal Product

As an article by Leander Kahney in *Wired* reveals, the stimulus to enter the digital music player category came from Apple's need to sell more iMacs.[4] Realizing that digital music was hot, Apple licensed existing music playing software and refined it. iTunes was then launched at the MacWorld expo in January 2001, providing Mac users an easy way to store and play their music.

In the process of developing iTunes, Apple realized that no good personal digital music player existed. Flash-based MP3 players had been around for a couple of years but had very limited storage. Hard drive–based music players were big and clunky. Jon Rubinstein, the engineer charged with developing what would become the iPod, realized that a new hard drive developed by Fujitsu that was only 1.8 inches in diameter could form the basis of the new product.

To bring the iPod to market in a limited time frame, Apple leveraged internal and external expertise, standard parts, and custom workarounds, but ultimately it was the design of the user interface that won the day. The iPod was simple, seamless, and easy to use. Interviewed by Rob Walker for the *New York Times* in 2003, Jonathan Ive, Apple's vice president of industrial design, stated: "What's interesting is that out of that simplicity, and almost that unashamed sense of simplicity, and expressing it, came a very different product. But difference wasn't the goal. It's actually very easy to create a different thing. What was exciting is starting to realize that its difference was really a consequence of this quest to make it a very simple thing."[5]

Apple not only created something different, it created an archetype for all digital devices. It became the standard by which all the alternatives were judged, but that did not stop other technology companies from trying to go one better. Philips, Samsung, and Dell all launched competitive devices, but none were as successful as the iPod. As Rob Walker noted, although the Dell DJ was slightly larger than the iPod, it held more songs and claimed a longer battery life, "which the company says is what its consumer research indicated people wanted."[6] Walker also reported that Steve Jobs dismissed the competition from Dell on the ground that they did not think about design or spend money on it. Jobs was justified in his belief that simple, intuitive, well-designed technology would command a price premium and sell well. Dell's sales were disappointing, and when it ceased production of the DJ in 2006, its U.S. market share sat below 3 percent.[7] The fundamental reason for Dell's failure was that it was trying to make a better product—not create a better experience.

2. Design Evolution

Technology is a fast-paced business, and you cannot afford to rest on your laurels. By the time Rob Walker was writing his *New York Times* article on the iPod in November 2003, two years after the product was first launched, the iPod had been updated twice and came in three memory sizes, all of which improved on the original's songs-to-space ratio. Soon after, the iPod Mini was introduced. A year later, in January 2005, the iPod Shuffle was introduced (followed shortly thereafter by a new generation of the Mini). And so it continued. As of the time of this writing, the Nano, which replaced the Mini, is now in its seventh generation. The Shuffle is in its fourth generation. The iPod Touch and iPhone are both in their fifth generation. Each generation has sought to improve on the previous one in some tangible way, although whether the changes will continue to be more than incremental is up for debate. What is fairly certain is that not long after this book is published, a new generation will have been added to each of the models listed previously, assuming they have not been replaced altogether.

3. Ecosystem

iTunes is now in its eleventh generation and, apart from simplicity of product design, is possibly the single biggest reason for Apple's success. The launch of iTunes challenged Silicon Valley's belief in interoperability. Normally, software is designed to work across multiple platforms, but an iPod would not play file formats that were rivals to iTunes, and downloads from iTunes would not play on rival devices. Strange as this move seemed at the time, it has yielded dividends. iTunes has become the glue that helps keep people loyal to the Apple brand. The more songs you download, the less likely you are to opt for a competitive system.

Of course, this advantage has not gone unnoticed, so today we see the battle of the ecosystems play out as Apple, Google, Microsoft, and Amazon fight for control of our digital world. But creating an ecosystem is not the purview of technology companies. Take Nike as an example. The brand started by making running shoes, but now service has become an integral part of the brand experience. Nike has built a whole ecosystem based around runners—from performance measurement with Nike+ to portals for advice and support online. NIKEiD provides customization services in store, online, and via apps like PHOTOiD.

While I see these three reasons to be the core of why Apple has remained so successful, there are multiple layers to the brand. At the heart of those layers is the design, both functional and aesthetic. Then there is the story of Apple, so long bound up with Steve Jobs, and the identity that the brand projects into the marketplace through its actions and the simplicity and tone of its marketing communications. The organization is aligned to deliver an approachable and enjoyable brand experience.

When I started writing this chapter, I hesitated to use the example of Apple. The brand and its story are simply so well known. But then again, it is tough to avoid referencing the most obvious example of a brand that has created enormous value through its single-minded focus on providing a meaningful user experience. To echo Ive, Apple products are simple by design, and different as a result. As a technology brand, Apple has a significant advantage over many brands that exist today. Technology is developing and changing all the time. While this constant change presents both an opportunity and a challenge, it does mean that the potential for tangible differentiation is that much greater. What about brands that are operating in a largely undifferentiated and commoditized product category?

MAKING A DIFFERENCE IN AN UNDIFFERENTIATED CATEGORY

Innovation is the main mechanism by which brands can create differentiation in commoditized categories. By improving people's product experience, brands improve the probability that they will attract new customers and retain existing ones, but product innovation alone is not the key to growth. Brands still need to find a way to communicate the news in a way that resonates with the target audience. Sometimes the difference is not really the innovation itself but just the way it allows you to tell a new story and align your brand experience with that story.

Roshfrans is a motor oil company founded in Mexico City over fifty years ago. In 2008, the brand was widely known and the fourth-largest lubricating oil brand in Mexico. Compared to its American and European competition, Roshfrans was more likely to be perceived as traditional and widely available, but it lacked any compelling functional or emotional benefit. In fact, research conducted by Millward Brown found there was little differentiation in the market in general. All the brands claimed to protect a car's engine, but none of them had substantiated how their product did so.

Product development offered a means for Roshfrans to formulate a rationale for better protection and the opportunity to set itself apart from the competition. After a prolonged development process, Roshfrans created its Ti-22 engine oil. The oil contained titanium, represented in the name by its periodic table symbol, Ti, and its atomic number, 22 (Ti-22). The new ingredient offered significant benefits: it refilled damaged surfaces and created a protective layer on components subject to friction, resulting in less wear, lower gasoline consumption, and lower gas emissions. The question was how best to communicate these benefits?

In order to launch the new product, Roshfrans developed new packaging and advertising. The new packaging was a contoured bottle, colored silver to signal the new ingredient. In brand tracking research, the majority of respondents preferred the bottle's look to that of the key competitors. In partnership with Grey Advertising, new TV spots were developed to support the launch with a new claim, "Friendly technology within everyone's reach," along with radio and billboard ads.

Two different creative approaches were used, supported by a 50 percent increase in the brand's SOV over 2008, making Roshfrans the biggest advertiser in the category. The functional advertising showed the oil as a silver liquid that flowed from the new bottle and coated the engine in a silver sheen. The ad described the various benefits offered by the new product. This fairly typical dramatization of the product benefits was accompanied by a more radical approach: one that sought to position the car as an extension of its owner. The basic idea was that if your car is in good shape, then so are you. This message was conveyed through a series of humorous executions, including one in which the driver transformed into a superhero.

The launch was successful. By amplifying the benefits of the new ingredient, Roshfrans had created a new and differentiating reason to choose the brand. The perception that the brand was high technology improved substantially, and a Millward Brown study showed the proportion of people claiming that Roshfrans was their first-choice brand rose to 25 percent in 2010. As a result of the launch, sales of Roshfrans grew by 33 percent in 2009 and by a further 11 percent in 2010.

If you can create differentiation through some tangible feature or quality that your target audience finds meaningful, then do so. Amplify news of that differentiation as loudly as possible. Make sure that people associate that benefit with your brand—not the category—and make it your own. If you do the job right, then that benefit will remain strongly associated with your brand long after the competition levels the playing field. However, recognize that although this approach offers the opportunity to make a step change in market share, it is difficult to sustain. By relying solely on innovation to grow your brand, you commit the brand team to a treadmill on which they have to keep up a constant stream of meaningful innovation. If you are not careful, the stream of "innovation" can become a stream of meaningless line extensions that undermine brand equity rather than promote it.

Creating meaningful innovation is not the only means of sustaining growth; in fact, plenty of evidence suggests that brands in undifferentiated product categories can still grow by creating intangible differentiation and amplifying that differentiation on a continuing basis.

BEYOND TANGIBLE DIFFERENTIATION

With lots of good-quality brands and plenty of copycats, people are spoiled with choices. Focusing solely on functional benefits can lock a brand into a comparative strategy—"Our brand is better than the alternatives because of X, Y, or Z." This tendency can be avoided if the brand can be made to "own" a particular functional benefit in people's minds, but a consumer-oriented goal-based strategy allows a brand to say, "We help you achieve what you want." So brands need to create demand based on noncompetitive factors, or, to use gaming parlance, they need to "level up" to create new origins for meaningful differentiation. However, the "territory" that a brand seeks to claim for itself needs to be consistent with the brand's purpose and delivery and build upon those factors.

I believe that the reason we see more efforts to go beyond physical differentiation has less to do with the consumer (who has always valued things that are meaningful) and more to do with the world of business and marketing. With oversupply in many categories and fast followers everywhere, functional differentiation is rarely sustainable. You have to take it to the next level to ensure that your brand stands out from the crowd, and that means seizing the high ground first. When the category is undifferentiated, brands can still create differentiation by creating links to something that any brand in the category could have done.

In the following pages, I list some of the more common approaches to leveling up. None of these are new. They tap into basic human motivations and have been used by savvy marketers for decades. Importantly, you will note from many of the examples cited that brands often use a layered approach that helps reinforce their differentiation.

Claiming Status

Gillette has become one of the most powerful and well-differentiated brands in the world by claiming the status of "The Best a Man Can Get." The only, the first, the favorite, the best, the most popular—whatever the specific claim, the idea that the brand is a leader in some way will create the perception that the brand is a safe choice. Depending on the culture, category, and individual, the allure of this type of differentiation will vary, but it is nevertheless a powerful influence that appeals to our need to simplify decision-making and go with the herd.

Individual Emotional Reward

Whether it is happiness, confidence, pride, satisfaction, or contentment, brands can identify themselves with specific feelings that consumers desire and that are appropriate for their product category. After struggling to recapture the sense of positivism embodied in its historic 1971 "Hilltop" TV spot, Coca-Cola has successfully gained the high ground of "happiness" in its "Open Happiness" campaign.

Self-Confidence

Often brands seek to address specific anxieties and concerns. James Young, a copywriter with J. Walter Thompson Company, helped launch the antiperspirant category when he presented perspiration as a previously unspoken social faux pas. Although his 1919 print ad, which suggested that a woman's stinky underarms might cause her to lose her man, met with outrage, sales of the advertised brand rose 112 percent the following year.[8]

Self-Esteem

This approach focuses on the feeling of self-worth enjoyed by the user (and, by implication, their difference from others). For example, one person to whom I spoke recently told me that she feels "righteous" when she drives her Toyota Prius. Dove's Real Beauty campaign successfully taps into women's desire for self-esteem in the United States and Europe, although it has been less successful in countries like China, where beauty is still desired in its own right. L'Oreal has leveraged the concept of self-esteem for many years but has adapted its slogan from "Because I'm worth it" to "Because you're worth it" and, more recently, "Because we're worth it" in order to stay in touch with the cultural ethos of its target audience.

Self-Expression

Apple famously appeals to people who "think different" and, by implication, suggests to the user that he or she is different, cooler, and more hip than people using a Dell, HP, or Lenovo. So the brand is used by consumers to signal "This is who I am." But we also use brands to signify achievement (think Montblanc or Mercedes), allegiance (think Manchester United or Patagonia), and independence (think Jack Daniel's or Harley Davidson).

Cultural Empathy

For a long time in the United States, Pabst Blue Ribbon beer was out of favor with the mainstream. Cheap and readily available, it was an every-man beer that no longer appealed to every man. But by appealing to the anti-establishment views of the bike courier community, Pabst managed to make itself meaningful to a very specific audience. While espousing a specific audience creates a perception of differentiation, there is a downside to this strategy: not everyone will want to be associated with that audience.

Tone of Voice

Any brand can choose a cheeky, fun voice for its marketing communications, and many have. But in the United Kingdom, Innocent Smoothies has created a well-differentiated brand on the basis of its projected personality and tone of voice. Combined with the brand's promise of 100 percent pure fruit, no sugar, no concentrates, and no funny business, the brand's distinctive tone of voice has helped it become successful.

Distinctive Icon

Small, green, and with a London Cockney accent, the GEICO Gecko first appeared in 2000 and soon became an advertising icon. Insurance stands out in our BrandZ research as being one of the categories most driven by brand. The majority of people think that their choice of a company matters, even if they do not have any preconceptions about which company is best. The challenge for the advertiser then becomes one of establishing some way to stand out from the crowd. To this end, the Gecko provides the means to deliver the single-minded message that GEICO makes buying insurance easy by saving you time and money. The advertising also helps give the brand a fun, outgoing persona that stands out from the other major players, who are seen as sensible and thorough.

Provenance

Cheese, cars, and champagne all benefit from being sourced from a specific region or country. Whether it is a specialty, skill, or quality, the associated image is usually built up over time, and the challenge is to interpret that quality in a way that is specific to the individual brand and not just the category. Toblerone, the iconic chocolate bar shaped like a range of mountains, makes reference to its Swiss origins and compares its shape to that of the Alps (even though the bar was supposedly inspired by a line of dancers at the Folies Bergères in Paris).

Whereas the preceding examples are all somewhat abstract in nature, some brands have found ways to level up by creating more tangible value for their consumers.

Societal Contribution

The Co-op, Kohler and Unilever, separately built brands and made an important societal contribution by actively seeking to improve the lives of their workers, and, as detailed in Jim Stengel's *Grow*, this form of competitive advantage is becoming more widespread today. Chipotle is committed to "food with integrity," seeking out the best ingredients produced with respect to animals, the environment, and farmers. Seventh Generation aims to respect the needs of the next seven generations in the way it produces and distributes its environmentally friendly home care products.

Special Experience

Special events have long been a way that brands create a unique experience for their consumers. For twenty-odd years up until 2009 American Express sponsored tennis at The Queen's Club Championship in London and gave free tickets out to cardholders (usually via competitions). Red Bull hosts a wide range of adrenaline-fueled events around the world that are attended by millions of people each year.

Brand Ecosystem

Today Apple, Google, Microsoft, and Amazon are locked in a battle for control of our digital world. In addition to their heartland businesses, they each now offer an ecosystem of mobile devices, apps, and cloud storage that work better with other items in the brand's ecosystem than with the same item offered by other brands.

My ROI

As noted earlier in this book, not everyone values brands, so there is always a niche for brands to offer an adequate, affordable product. In some cases, customers may hate the service but still use it. Ryanair in Europe is a classic example. Here, the value of the brand is the money saved by the purchaser that can be directed to other needs.

Personalization or Mass Customization

Nike and niche automotive brands like Mini and Scion have made personalization an important part of their offers. Scion is the progeny of Toyota, and for the most part, its cars are unusual body shells grafted onto the standard Toyota chassis. But if you live outside North America, you won't recognize the name Scion, because it does not exist elsewhere. Scion was created specifically to target younger consumers in North America. The Scion brand is expressly designed to inspire individuality and allow owners to express their unique personality by customizing their cars. No two Scions are intended to be alike. An *AdAge* video reports that Scion owners spend twice as much on customization as they do on the original car.[9]

Customer Commitment

Avis's slogan "We try harder" was originally coined in 1962 as "We're No. 2. We Try Harder." Positive customer experience ratified the claim and helped Avis achieve significant sales growth. Fifty years later, this signal of commitment has been replaced by "It's Your Space,"' which lacks the meaningful differentiation of its predecessor. Zappos places the joy of the customer above the need to sell him or her anything (and in the process sells vast quantities at a full profit margin).

Of course, none of these strategies is mutually exclusive. Most brands would be well advised to create an integrated set of mechanisms by which they create meaning—provided these mechanisms are recognized and understood to be synergistic and to have a common origin—rather than simply diluting what the brand stands for in people's minds.

WHO'S YOUR COMPETITION?

For an established and successful brand, it is very easy to get locked into a particular mindset and to make assumptions about what brands form the competitive set. However, some of the most disruptive and challenging competition comes from outside the product category. First, it is easy to ignore such a threat, because it is probably not measured, and, second, this threat can be difficult to counter when the alternatives trespass on what used to be your brand's differentiating benefit.

Faced with a drop in sales, in 2009, Clorox Bleach launched a campaign to illustrate the different ways bleach could be used around the house. Until then, the brand had typically focused on its core benefit: helping to make clothes whiter than could laundry detergent alone. Interviewed for *Adweek*, Clorox Bleach marketing manager Chris Hyder notes, "The marketing insight is, even though this brand is [almost] 100 years old, in many ways, it has to be treated like a new product. There is such a lack of understanding of the many benefits of bleach."[10]

This change actually does represent a return to the brand's origins— Clorox was initially marketed in 1916 as a "bleacher, germicide, cleanser and disinfectant"—and points to the fact that what made a brand successful in the past may help reinvigorate it today.

Normally, this approach might have helped expand category usage, but unfortunately, the issue facing Clorox today is more fundamental than simply making people aware of new uses for the product. According to *Euromonitor*, U.S. consumers have become increasingly uncomfortable about using bleach, and sales have steadily declined over the last five years.[11] Additionally, innovation in adjacent categories, notably laundry and surface care, has provided consumers with more convenient sanitizing options than bleach. So the task facing Clorox is far larger than simply expanding market share (it is already the dominant bleach brand and did gain share against store brands following the 2009 campaign). This example points to the importance of knowing the brands against which you are competing. Clorox might be a strong brand in the bleach category, but it is weaker than Tide, against whose stain release product it must now compete.

HOW DO *YOU* DIFFERENTIATE?

In Chapter 2, we explored how a difference—no matter how slight—can make it easier for people to choose your brand over another; in Chapter 4, we explored the importance of perceived differentiation to a brand's ability to both command a price premium and grow market share. If, as some pundits would have us believe, differentiation is dead, then so too is our ability to create strong and successful brands. True product differentiation may be hard to achieve, but there are many ways to create differentiation. The question is, which of the myriad options is going to be right for your brand?

Questions to Consider

1. Is your brand different enough? Map your market using different techniques and different dimensions to highlight how closely brands cluster together and whether some are distinguished by particular attributes. In any market, a consumer will be able to say relatively easily whether he or she perceives a brand to be different from the rest, but you might need to dig deeper to understand the underlying reasons for that perception. One aspect that must be treated with care is the size of the brand. Dominant brands can appear to be less differentiated because they effectively own the category. Is there a hidden difference that consumers have forgotten? Interview people at the point of purchase to find it.

2. Can you identify ways to create and then improve your differentiation? There are two traps to avoid. Do not try to differentiate on basic category requirements unless you really do have a better mousetrap. Equally, it will be easy to differentiate your brand using less relevant attributes, but this approach might not be compelling enough. A marketer will need to judge from his or her own experience what the important levers are that could genuinely accentuate the brand's difference, either through the way the brand is presented in the market place or by changing the nature of the offer itself. Remember, it may be possible to seize an unclaimed generic benefit for your brand or make the unimportant important. Try to predict the potential competitive response to each potential strategy.

Chapter 9

Defining a Meaningfully Different Experience

The four factors outlined in the preceding chapters—purpose, delivery, resonance, and differentiation—combine to create a brand's meaningfully different experience. This is not a sequential process. For instance, your brand's purpose will inform the audience to whom the brand appeals, what is most important in terms of delivery, and what will differentiate your brand from the alternatives. Identifying a brand's meaningfully different experience is not linear; it is a matter of synthesis. The fusion of the four key factors is what determines whether or not a brand is meaningfully different.

The successful definition of what makes a brand meaningfully different is the tipping point in our ValueDrivers workshop. We have yet to conduct a workshop in which there was complete clarity regarding the brand's purpose, let alone regarding what made it meaningfully different, but without that definition, the remainder of the workshop, which is focused on successful amplification, is useless. The critical nature of the definition makes this section of the workshop very stressful for both the participants and the facilitators.

In the workshop, we have the participants work in groups. Each group has completed a series of exercises designed to examine the brand's purpose, delivery, resonance, and competitive standing. By the time the participants have to define the brand's meaningfully different experience, they have been thinking about the brand from different perspectives for over three hours, and it is time to pull together everything they have discussed in one succinct statement. We are not looking for a tagline, but for something that captures the essence of the brand. We give participants twenty minutes to complete the exercise and provide them with the following instructions:

Craft the meaningfully different experience for your brand in just a few words (four to eight!). The meaningful difference

- *Must be backed by the brand's purpose*
- *Reflected in the brand experience*
- *Informed by what is differentiating about the brand*
- *Defined in a way that resonates with consumers*

While this activity is probably the most intense section of the workshop, with people working under pressure to complete a relatively complex task, we always end up with a set of well-crafted concepts that usually display remarkable consistency across the groups. Providing a framework for exploring the brand and allowing time for feedback on and synthesis of the output from the exercises invariably leads to a more consistent viewpoint of what the brand really stands for. This definition acts as the connection point between what the brand does and how people respond to that experience.

One of the questions that often comes up during this session is, "Are we defining what the brand stands for now or what we would like it to stand for?" The answer is both. You have to assume that you are starting from where your brand stands today. Too big a stretch will leave your customers and consumers floundering.

Obviously, I cannot share the output of specific client engagements in this book. So in order to give you an idea of how things work, let us examine one of today's well-known brands and see if we can identify what makes it meaningfully different.

◼◼◼◼ THE BIRTH OF A MEANINGFULLY DIFFERENT BRAND

A while ago, I came across an article originally published in the August 17, 1981, edition of *People* magazine that described how Häagen-Dazs came to be.[1] What initially caught my attention was that the Häagen-Dazs name was made up—it does not mean anything. But as I read on, I realized that the article was describing the birth of a meaningfully different brand, one I could easily map using the ValueDrivers framework.

Purpose: Providing a Premium Ice Cream for Adults

Launched in 1960, Häagen-Dazs was a brand born of necessity. Founder Reuben Mattus, a Polish immigrant living in New York's Bronx borough, had been making ice cream for years, but in the 1950s, the large mass-producers of ice cream started a price war. Mattus reports, "I realized I couldn't keep up and maintain any kind of quality. I thought maybe if I made the very best ice cream, people would be willing to pay for it."[2]

Up until this time, ice cream had typically been regarded as a treat for kids, but if Mattus was going to be successful in selling a premium product, then he would need to break the mold and target adults.

Delivery: Less Air and Premium Ingredients

Mattus's competitors were able to drive down prices by pumping more air into their product, a process known as *overrun*. As detailed in the *People* article, Bassett's had a 75 percent overrun and Breyer's a 92 percent overrun, but Häagen-Dazs's was only 20 percent. The lower overrun gave Häagen-Dazs a richer taste and a creamier texture. The product had no preservatives, additives, or stabilizers and was made only with fresh cream, milk, and eggs—a recipe that still holds true for the brand today. In order to deliver the best taste, Mattus sourced premium ingredients like Madagascan vanilla beans, South American coffee beans, and real rum for its rum raisin flavor.

Resonance: Great Experience and Exclusiveness

The Häagen-Dazs experience resonated with consumers, and word of mouth helped spread the news. People loved the ice cream in its own right, but the premium positioning of the product added to the brand's appeal with what the *People* article refers to as "a healthy dash of snob appeal."[3]

Difference: Richer-Tasting and Upmarket

The richness and creaminess of the product is the foundation of Häagen-Dazs's success. Mattus's daughter Doris is quoted by *People* as saying: "Our customers are repeaters. Once they taste Häagen-Dazs, we've got them for life."[4]

At first, the company sold just three flavors—chocolate, vanilla, and coffee—and they were sold only in gourmet shops around Manhattan, which added to the brand's cachet. And then there's the name. While Häagen-Dazs does not mean anything literally, the name does carry implicit meaning, just as Reuben Mattus intended when he and his wife Rose started brainstorming names around their kitchen table. In order to differentiate his brand from cheaper competitors, Mattus wanted people to think that the brand was imported, and he thought the Häagen-Dazs combination would confer an association with Denmark's well-known dairy business.

Justifying a Price Premium

What Reuben Mattus did in creating the Häagen-Dazs brand was to create a meaningfully different experience, one that had not been available in the ice cream category before. His counterintuitive response to a price war was to create an expensive brand, one that was clearly differentiated from the existing brands in terms of product, target audience, and positioning. When Häagen-Dazs launched, its price was 40 percent higher than the competition's, but that did not stop the brand from becoming successful. At the time the *People* article was written in 1981, Häagen-Dazs was expected to sell 6 million gallons, or $40 million dollars worth of ice cream (nearly $100 million in today's money), over the course of the year.[5]

Leveraging a Strong Brand

Once created, a strong brand becomes a gold mine if handled right. Simply extending the brand's footprint to new markets can multiply its earnings significantly. In 1981, Häagen-Dazs had a booming franchise business, with more than 92 stores nationwide. Fifty years after its creation, Häagen-Dazs is sold in more than 80 countries outside of North America by General Mills in partnership with Nestlé. While the brand's super-premium positioning dented growth during the Great Recession, sales returned to an annual growth rate of 7 percent in 2011, contributing $750 million in sales to General Mills's income.[6]

██████ A MEANINGFULLY DIFFERENT EXPERIENCE STANDS THE TEST OF TIME

As the example of Häagen-Dazs demonstrates, a meaningfully different experience can stand the test of time. If you have correctly identified what resonates with your customers or consumers, then there is a good chance that it will continue to do so for decades—provided, that is, that the original idea does not get forgotten, corrupted, or swept out by a new broom. Many brands have learned, at their own cost, that changing what you stand for is far from easy.

In the late 1990s, the glitzy city of Las Vegas tried to shift gears and market itself as family-friendly. As Leo Hopf and William Welter note in their book *Rethink, Reinvent, Reposition*, Las Vegas's attempt to ignore the city's origins and heritage proved unsuccessful: "The city tried to hide or at least gloss over the gambling, drinking, and sexual freedom that had been a part of Las Vegas since its founding. On the whole, the campaign was a failure; families with small children did not come to the city in large numbers, and the idea of a family-friendly Vegas discouraged those in search of an adult-oriented playground."[7]

I remember thinking at the time that the idea was a stretch. Everything I had ever heard about Las Vegas seemed to be the contradiction of a kid-friendly destination. In 2002, Las Vegas admitted defeat and returned to a campaign more in line with its origins: "What happens in Vegas stays in Vegas."

This example highlights the fundamental problem facing most brands. It is simply not possible for a brand to be all things to all people. By seeking to attract new users with a different mindset from that of the existing target audience, you risk undermining the existing users' ideas about the brand. You may win over some new users, but in the process, the brand loses its appeal to existing ones. Brands pursuing this approach often end up undermining their ability to charge a price premium, because they have eroded the differentiation that justified that price point for the original target audience, while failing to establish credible meaning and differentiation for the new audience. The brand ends up trading on price.

There will be times when it is necessary to change positioning. Sometimes a trend will make a brand's meaningful difference redundant or less resonant. In the United Kingdom, the carbonated glucose drink Lucozade was marketed as a product that aided recovery from illness. When faced with a general improvement in the public health of the population, the brand successfully repositioned itself as an energy drink. Sometimes marketing a brand in a new cultural context will require change, but often this change manifests itself as a reinterpretation of what the brand stands for, not a complete shift in meaning.

HÄAGEN-DAZS IN CHINA

Häagen-Dazs may have been born in the United States, but today it faces strong competition in its homeland. Ben & Jerry's, the brainchild of two Vermont entrepreneurs, is another meaningfully different brand that has eroded Häagen-Dazs's standing in the U.S. market. Now owned by Unilever, Ben & Jerry's has successfully developed a strong alternative to Häagen-Dazs, offering flavor combinations with whimsical names like Chubby

Hubby, Americone Dream, and Phish Food (named for the cult Vermont rock band). If you want to see Häagen-Dazs at the height of its power today, you need to go to China. For an American familiar with finding pint tubs of Häagen-Dazs in every grocery store freezer cabinet, Häagen-Dazs in China offers a dramatically different experience, but one that nevertheless builds on the brand's strong heritage and upmarket credentials.

In China, Häagen-Dazs presents a unique, indulgent, and adult ice cream experience, primarily through its positioning in retail stores. Located in the ritziest parts of town (a strategy that harkens back to the very earliest days of the brand's growth in Manhattan), the brand's retail cafés offer a luxurious dining environment. Customers select from an extensive menu of desserts that are designed to be shared with loved ones and friends.

These concoctions are not simple scoops of ice cream with some nuts and hot fudge sauce. When I was last in Shanghai, the menu offered a special series of creations from some of the world's most famous chefs. One item, created by French pastry chef Jean-François Arnaud and labeled "Fashion in Paris," features an ice cream handbag and is described in the menu as follows: "Fashion in Paris, the delicate bag is made from Caramel Biscuit & Cream Ice Cream, the Strawberry Ice Cream wears a crispy hat, the white chocolate box is full of Chocolate, Summer Berries, Rum Raisin and Cookies & Cream Ice Cream, decorated with a strawberry macaroon."

The menu goes on to feature four more similarly exotic creations, including an ice cream violin and a representation of Cologne Cathedral, before itemizing its more mundane but nevertheless striking classic creations.

■■■■ MAKING A VIRTUE OUT OF NECESSITY

When General Mills launched Häagen-Dazs in China in 1996, they chose not to manufacture the brand locally. The decision was based on two important factors: refrigeration in many grocery stores was inadequate and good-quality local milk was expensive. The Häagen-Dazs product is made with all-natural ingredients and uses egg custard, not stabilizers, so it needs to be kept at a constant, very cold temperature, something that local grocery stores were unlikely to deliver. Similarly, concerns over the quality and safety of local milk supplies led to the decision to import the product from abroad. The same practice is still followed today. When I interviewed Gary Chu, CEO of General Mills China, he reported that they had recently reexamined the decision to manufacture abroad but had confirmed the original decision, in part because quality milk is far more expensive in China than elsewhere. Product is still flown in from France and sold either in the Häagen-Dazs stores or from its own grocery store freezer cases.

The decision to import product imposes significant transport costs and import duties that must be recovered at the point of sale. The indulgent dining experience, extravagant desserts, and upmarket locations of Häagen-Dazs are all designed to justify a super-premium price point.

◼◼◼◗ WHAT WILL APPEAL TO NATALIE?

Bear in mind that China has no dessert tradition, so to be successful, General Mills had to introduce consumers to a new experience and do it in a way that would appeal to Chinese sensitivities and tastes.

Chu reports that when General Mills opened the first Häagen-Dazs café in Shanghai in 1996, the team had no firm plan for positioning the brand but used the shop as a chance to introduce Häagen-Dazs to the Chinese and see how they reacted. By 2001, General Mills had invested $10 million in the expansion and were busy refining the brand's positioning. They discovered that the indulgent experience appealed more to women than to men, and so they set about characterizing their target consumer in terms of lifestyle, attitude, and interests. Through extensive consumer research, they built up a picture of a person they called Natalie.

Natalie is 30 and single. She is not defined by income but by lifestyle. She likes the quality life and enjoys spending time with friends, but she also values her private moments and indulgences. The identification of Natalie led to the development of the tagline "Love yourself, love Häagen-Dazs" and informed the brand strategy on everything from store locations—where would Natalie go?—to the emotional nature of the brand positioning. The Häagen-Dazs team works hard to keep up with the changes in Natalie's life as it evolves in response to China's fast-moving culture and to adapt their approach accordingly.

Chu states, "Emotional bonding is most important, especially in China." Romantic love has become a central theme to the Häagen-Dazs brand in China. He further notes: "Love is universal—the origin of so many stories. We have had people getting engaged in the stores, and the stores feature in movies as a romantic setting. And status is important. People want to sit by the window to be seen and show off, safe in the knowledge that they are sitting with the same class of people."[8]

The important thing is that, while love is not inherently associated with ice cream, the sharing experience offered by the café means that the emotion is rooted in the brand experience, not just layered on or tangential to it.

FINANCIAL VALUE GROWTH

Fortune magazine estimated that in 2010, Häagen-Dazs brought in revenues of over $100 million in China and grew at an annualized rate of 21 percent over the three previous years.[9] Currently, the majority of that revenue comes from ice cream parlor shops as opposed to supermarket sales. In our meeting, Chu stated that he expected even higher year-on-year growth in 2012 from the brand's associated stores, gifts, and food services. He said, "Our real money-making is in our creativity." The company plans to introduce 20,000 more freezers over the next three years, and as Chu noted, with a current presence in 37 Chinese cities, there are at least another 300 to go. But Chu believes in not rushing things. "Branding is forever," he said. "I advise my guys: patience. We are not just here for one year, we are here for the long term."[10]

MEANINGFULLY DIFFERENT EXPERIENCE IS DURABLE

I hope the example of Häagen-Dazs gives you some idea of how a meaningfully different experience can be defined based on the brand's purpose, delivery, resonance, and difference. For lack of real client examples that I can share, I have had to "reverse-engineer" the four elements based on what I know about Häagen-Dazs's origins. Did Ruben Mattus consciously consider all of these elements when he created the brand in 1960? Maybe not. But he did know that in order to remain in business, he was going to have to innovate, to do things differently. He was not prepared to make a cheaper product, so he had to find ways by which he could justify a premium price, and the way he did so conforms to our framework.

Once identified, a meaningfully different experience proves remarkably durable. Häagen-Dazs's recent success in China illustrates that although the execution might be tailored to a new culture, the original essence of what made the brand successful remains: an unwillingness to compromise on quality and an indulgent product experience. Knowing that essence and the motivations of the target audience has allowed Häagen-Dazs to thrive in a very different culture from the one in which it first found success. The brand's meaningfully different experience informs how the brand goes to market and makes it easy to decide what to do and what not to do.

Questions to Consider

1. Can the brand owners articulate in a single sentence what the meaningfully different experience is for the brand? Is there a shared agreement that opens up possibilities and makes the team excited about the future of the brand? Is this experience something that no other brand could own?

2. Do the brand owners have the same vision of the brand's meaningfully different experience? If a consensus cannot be reached, then the elements of purpose, delivery, resonance, and difference must be revisited, re-explored, and shared among the team until such consensus can be built. The meaningful difference should be attainable but challenging, something to which the brand could aspire in the short to medium term, and it should be clear what needs to be done for this end to be attained.

PART III

AMPLIFYING A MEANINGFULLY DIFFERENT BRAND

Faced with retail and media channels that are proliferating and fragmenting simultaneously, marketers understandably feel challenged by the task of implementation. Which new retail channels present the best opportunity? How do we distribute our brand in India's rural markets? Which media channels do we need to use? How do we best promote our brand though social media? In discussing the multiplication of channels and touch points at which marketers must exercise creativity, Tom Buday, Nestlé's head of marketing and consumer communication, states, "The nature and breadth of creativity required from marketers today is mind boggling."[1]

So where to focus? Buday suggests that a brand's "impact ambition" should inform all its actions. Nestlé's impact ambitions are short, succinct impact statements of the difference the brand is intended to make in the lives of the people it serves. It is a philosophy that is very similar to the one proposed in this book. If you know what makes your brand meaningfully different, then you should seek to reflect that quality in all aspects of the consumer experience. The meaningfully different experience would thus become a reference point for all decisions made about when, where, and how to market the brand. The fundamental objective should be to ensure that the brand's meaningful difference remains apparent and salient across all touch points. Marketers can then use that difference to increase the financial value of the brand.

The following chapters will address the five primary ways in which marketers can generate increased value; each chapter will provide examples and food for thought. These five ways offer methods for amplifying your brand's presence that have been tried and proven repeatedly.

Findability is the foundation on which a brand is built. Unless your brand is easily found when and where people want it, then it will never realize its full potential.

Credibility is the process of enhancing a brand's meaningful differentiation through innovation and association with brands, people, and events.

Vitality refers to the need for a brand to be seen as contemporary and thriving. Nothing succeeds like success, and people are drawn to brands that are seen as popular and are talked about.

Affordability details the uneasy relationship between brand value and price. Manipulating a brand's price is the easiest way to generate additional volume sales and, at the same time, the biggest threat to sustainable, long-term profitability. We will explore how brands can work with a specific price point and still achieve long-term value growth.

Extendability explores the biggest growth driver of all. Once you have created a strong, meaningful, and different brand, extension into new categories and countries offers a way to maximize the brand's value. Handled well, brand extension makes a strong brand an incredibly valuable asset.

Chapter 10

Findability

Findability is all about ensuring that your brand is available and easily found when and where people want it. If this sounds like something you learned in a marketing basics class, that's because it is—but that does not make it any less important. In fact, findability is such a basic concept that people often take it for granted. They make assumptions about where the brand should be distributed. They overlook the need for the brand to be easily searchable—both digitally and physically. They fail to ask simple, necessary questions, such as: How can I make my brand more readily available? Can I gain new buyers by identifying new purchase occasions and situations that are not served by the competition? How can I make my brand the most obvious choice?

Physical distribution is the most important factor in determining the success of a new consumer packaged good; the second is brand awareness. These two factors work hand in hand. Seeing a new brand stocked in a store does not guarantee that someone will pay attention to it or want to check it out. Most of the time people already have a brand or set of brands in mind that they consider acceptable. Why bother with a new one? But if the marketer has done his or her job well, then the sight of the brand will evoke a strong positive impression and make people pause for thought, perhaps remembering, "Hey! That's the organic one/the one that kills 99 per cent of germs/the one that my son wants to try."

Findability is by no means limited to consumer packaged goods. If you sell a physical product, then physical availability is an obvious prerequisite to brand success. Without adequate distribution, a new brand will fail and an established brand will fail to maximize its potential. But even if you are selling your brand online, findability is important. Why would someone click on that link if the name means nothing to them? Two of the biggest catalysts for online search are offline advertising and word of mouth. These make a memorable and easily identified name and logo worth their weight in gold. How else are people going to find your brand? If people are forced to do generic category searches, likely as not, your competitors' names will come up too.

Within the ValueDrivers framework, we identify two key aspects of findability: availability and visibility. Both concepts are important, whether your brand is a physical product or an intangible service.

◼◼◼◼ AVAILABILITY

Physical availability of a retail outlet has obvious ramifications for consumer demand in any country. Across a wide selection of product categories measured in BrandZ, around one in four people agree that that their purchase decision is influenced by location, placing convenience over brand choice. For grocery stores, for example, this influence determines a large number of purchases, ranging from 20 percent in a well-developed retail market like the United Kingdom to as high as 55 percent in Thailand. Convenience not only dictates which grocery store many people shop at but also influences the demand for consumer packaged goods brands stocked in those stores. The same constraint can apply to financial institutions, car dealerships, clothing brands, fast food retailers, and more. Growing availability grows revenues. Based on data taken from Starbucks's annual reports between 1995 and 2011, a Millward Brown analysis found that every 10 percent rise in the number of Starbucks stores saw a 6.4 percent increase in revenues the following year.

If your brand is not easily available when and where people want to buy it, then you are missing out on sales. Identify the gaps in your distribution and work to fix them. Marketers should also look for opportunities beyond traditional distribution channels to open up new sales potential for their brands. All too often, we work within defined distribution channels: value-added resellers, grocery stores, online, and so on. The questions we need to ask are: Where else might people want or need my brand? What new distribution channels might potential brand buyers be using?

In a paper titled, "What Makes Big Brands Stay Big?" Craig Twyford, vice president of product leadership at Nielsen, makes an interesting suggestion. Examining U.K. brands that have stood the test of time since 1949, he suggests that these are the brands that dominated the newly emerging retail chains.[1] Twyford notes that Heinz baby food, Colgate toothpaste, and Wisdom toothbrushes, which were stocked in the developing supermarket chains, have stood the test of time, but their competition—Brands baby food, Kolynos toothpaste, and Spa toothbrushes, which were stocked in chemists and independent stores—did not. Twyford acknowledges that there is not enough data from 1949 to make an empirical conclusion but states, "Spending time working out which channels and which retailers will emerge as the leading force and focusing resources there is likely to pay dividends in the long term."[2]

While his conclusion is not a proven fact, many of today's most sophisticated marketers aim to ensure that their brands are readily available in new retail channels. In a presentation given to investors, Harish Manwani, COO for Unilever, talked about the need to grow in "breakout" channels: winning market share in drug stores, building scale in e-commerce, and accelerating the growth of new out-of-home retail opportunities.[3]

VIRTUAL SHOPPING

Chances are that when you think of consumer packaged goods, you think of the grocery store or your local convenience store. But the opportunity for e-commerce has not gone unnoticed by the consumer packaged goods companies. Although early attempts to sell packaged goods online met with little success, improved distribution systems mean that many companies are now trying again. But when shoppers are mobile and mobile-enabled, why have them go online to a website to buy your brand? Several different companies around the world are experimenting with mobile shopping from virtual stores.

Virtual subway shopping is the digital equivalent of the street vendors hawking their wares between lines of cars backed up on busy roads in Accra, Delhi, or São Paulo. The store comes to where the shopper is, instead of the shopper coming to the store. The basic concept is simple. Set up a virtual store with images of your brands in a busy subway station and allow consumers to order goods for later delivery using quick response (QR) codes.

While the idea of the virtual subway store might be simple, in 2011, Cheil Worldwide won the Grand Prix at Cannes in the media category for its "Homeplus Subway Virtual Store." Homeplus is Tesco's South Korean discount store chain, and the virtual store was proposed by Cheil as a way to bring the store to new customers. *Advertising Age* reports that although engagement with the store was measured in the tens of thousands, online sales for Homeplus jumped by 130 percent.[4]

The idea seems to be catching on. In the Czech Republic, Procter & Gamble set up four virtual stores in the busiest subway stations in Prague, selling brands like Ariel detergent, Gillette razors, Old Spice deodorant, Pantene shampoo, and Pampers diapers. And in Shanghai, China, where online shopping is huge and commuters seem to while away the time by staring at their mobile phones, online retailer Yihodian.com installed fifteen stores in local metro stations. In this case, shopping requires a Yihodian app downloaded to an iPhone or Android, but once that is done, simply scanning the bar code puts the brand into your electronic shopping cart.

But why limit your virtual store to the subway? Why not the sports stadium, airport, or movie theater? Anywhere where people congregate with time on their hands might offer an opportunity.

A Smart Acquisition Positions eBay to Stay Ahead of the Curve

It is indicative of eBay's success that in 2012, a woman's handbag was sold every 30 seconds on eBay's mobile platform.[5] Today, eBay's revenues are growing and the stock price is riding high. But only five years earlier, things were far less rosy. In 2007, eBay recorded its first loss as a public company with a $1.4 billion write-off on its $2.5 billion acquisition of the calling service Skype. But while Skype was an acquisition that had little to do with the company's online auction business, another acquisition proved to have everything to do with its future.

PayPal was acquired by eBay in 2002 for $1.5 billion and has proved to be a good long-term bet for the company. The online payment system has become integral to many websites and increasingly facilitates eBay's one-click payment solution for mobile. If there is something that mobile users would rather avoid, it is filling out lengthy forms. So a safe, simple, and convenient solution helps overcome any reticence about making a purchase from a smartphone. While eBay has expanded its operations from auctions into e-commerce, most of its growth now comes from mobile retailing. According to Kyle Spencer of the Seeking Alpha website, PayPal's revenue increased 26 percent year over year in 2012, driven primarily by increased penetration on eBay, as well as by continued merchant and consumer adoption and strong growth in Bill Me Later, which allows purchasers to pay for goods without using a credit card.[6]

Of course, the acquisition of PayPal alone does not explain eBay's current success. Credit also goes to CEO John Donahoe, who pushed the company hard in the direction of mobile. In a *New York Times* article, Donahoe states: "We saw the mobile revolution early and we made a big bet across the entire company. We saw that mobile was an important factor for our customers. It was becoming the central control

➡

device in their lives. We didn't worry if it cannibalized our existing business, because we knew it was what our customers wanted."[7]

That bet seems to have paid off, as eBay reported third-quarter revenues of $3.4 billion, up 15 percent, while profits climbed 14 percent to $718 million.

◼◼◼ VIRTUAL GOES PHYSICAL

These days you can buy pretty much anything you want online, from cars to carpets, cleaning products to clothes. But in an interesting juxtaposition, just as marketers of goods traditionally sold in physical stores are seeking to scale online, online-only retailers are discovering that it pays to set up physical stores, particularly when they satisfy people's desire to check out the product before buying it.

An article by Stephanie Clifford in the *New York Times* reports how Andy Dunn, founder of the e-commerce apparel company Bonobos has recently had a change of heart.[8] Dunn used to be an online-only evangelist trying to satisfy the needs of men who do not like the physical process of shopping for clothes. But now he offers them the opportunity to try before they buy in a physical store. Why? Because not everyone wants to buy sight unseen or order three sets of everything to make sure the sizing is right. But, as Dunn acknowledges, starting from scratch means you can design things very differently. The Bonobos Guideshop acts as the fitting room, and the website is its virtual back room, where orders are placed online and delivered to the customer. The results speak for themselves: the average in-store transaction for Bonobos is $360, double what it is online, and first-time store visitors generally buy again in 58 days, versus waiting 85 days between website purchases.

Of course, the downside of the Bonobos business model is that buyers have to wait for their purchases to be delivered. In Chapter 13, we will consider how SuitSupply is reinventing the business of selling men's suits by aggressively cutting the cost of a good suit; in addition, SuitSupply offers its buyers instant gratification by having tailors on site. Alterations are made immediately after a fitting so that customers leave the store with their new purchase in hand.

SuitSupply is not the only company to recognize that convenience is all well and good, but faster is better. Amazon is investing billions in warehousing and automated delivery systems in order to achieve the holy grail of same-day delivery. I suspect it will only be a matter of time before you find Amazon has opened a store on your local main street. After all, it would save you hassle and them the shipping costs if you could check things out before buying, right? Too radical an idea? Remember, not everyone is online or willing to buy online. Some people do like to check things out before they buy and dislike the hassle of returning stuff that does not live up to its promise. Amazon is a big brand with a strong reputation and a proven willingness to invest. The question is whether they can adapt their business model in order to make the economics work—not whether the opportunity exists.

EMERGING MARKETS

In a funny way, the drive to use the latest technology to make shopping easier and more convenient is simply a return to tried and tested retail formats. Amazon, eBay, and Taoboa are the modern-day equivalent of the traditional markets found in countries such as Italy, Spain, Kenya, and Chile, where vendors come together to sell their goods and produce. Virtual kiosks are not that different from the door-to-door salesman. The very same companies that are leveraging mobile technology to sell their brands in the big cities of the globe are also busy taking their brands to people in less-developed regions of the world.

In India, Hindustan Lever utilizes an army of around 45,000 "shakti women" to sell its products in remote villages. Project Shakti—a word that means "strength" or "power" in Hindi—started in 2000 and has now evolved to serve over 3 million households.[9] The project is not just about selling more Unilever products; Unilever is also providing microfinance and a source of income to people who would otherwise find it difficult to earn a living. Project Shakti has been so successful in India that Unilever is now extending the model to Nigeria and Kenya.

The big question for Unilever is how long this advantage in India will last. While Hindustan Lever is well established in India's existing retail ecosystem, modern trade, like supermarkets and hypermarkets, is beginning to gain ground, offering both an opportunity and a threat.

Modern trade could provide Unilever the opportunity to use its existing business base to dominate a new retail channel, or it could pose a threat because such trade levels the playing field and makes the price difference between brands that much more obvious. In the small, crowded, family-owned stores often found in India and Africa, people still need to ask for a brand by name or hunt it down in the store—they have to know what they are looking for. In a modern supermarket, however, a selection of brands is presented for sale with the prices displayed for eager shoppers to compare. Should the shopper be confused as to which brand to buy, there will often be a brand representative standing by to offer advice. Modern trade may make it easier for marketers to bring their goods to market, but it does not necessarily make it easier for them to keep their existing customers or make good margins.

VISIBILITY

Visibility encompasses everything that people associate with your brand, from its physical presence (if it has one) to its visual identity and searchability online. To make sure that your brand is as visible as possible, you must understand the sensory cues that help people recognize it and distinguish it from its competitors. You need to understand how to make your brand stand out on the street, in the store, or online and identify the cues that most easily trigger a positive response from those who do see it. Again, this is basic marketing, but have you mapped out all the touch points between a potential consumer and your brand in order to assess what will best differentiate it from the crowd?

My observations suggest that most marketers have not done so; otherwise, the way they present their brand to the world would not be so fragmented and inconsistent. Too often, marketers seem to assume that potential consumers will be looking for their brand, when, in reality, seeing the brand or its name is the trigger that stimulates people's interest based on previously dormant associations. This applies to both habitual and considered purchases. For instance, by the time someone gets to the car showroom, they may well have an active interest in a particular make and model, but in all likelihood, that interest was piqued by some random encounter with the brand elsewhere.

Automotive marketers have the advantage that the product and retailer are integrally linked, but for most other considered purchases, the first question that comes into people's heads might be "Where can I get the best deal?" rather than "Which brand do I buy?" Whether the best place to buy turns out to be a physical retailer or online, your brand really does need to sell itself. It needs to be distinctive and attractive.

HOW DO PEOPLE RECOGNIZE YOUR BRAND?

Do you know what cues lead people to recognize your brand? The evidence suggests that many marketers do not. Why else would they change the visual design of their brands in such a way that people fail to identify them? And yet that is what happens time and time again.

Perhaps the best-known packaging debacle of recent times was the short-lived redesign of PepsiCo's Tropicana Pure Premium brand. Tropicana is a well-known juice brand in the United States and has dominated the category for many years. In a relatively undifferentiated and habitual category, Tropicana was largely dependent on its salience for competitive advantage.

At the beginning of 2009, Tropicana launched a new pack design created by Peter Arnell, CEO of Omnicom's Arnell Group. As described by Arnell, the new packaging launch was the product of a five-month project to give "new energy" to the brand and evolve it into a "more current or modern state."[10] In a presentation recorded at a Pepsi press conference, Arnell described how the new design was intended to show the juice, not just the outside of the orange, and demonstrated the new squeeze cap, noting the comparison with squeezing a fresh orange and the less obvious parallel with the emotional connection between a mother and her kids.

Taken in isolation, the rationale for these changes might make sense. In fact, the idea that showing the juice was important was reflected in the success of Coca-Cola's Simply brand, which comes packaged in a see-through carafe. But Tropicana was not a new brand starting out with a clean slate. Tropicana's consumers were already imprinted on the brand; they recognized it by the iconic orange and straw on the front of the package. When that was replaced by a glass of orange juice, the pack became indistinguishable from the others around it. Worse, the redesign took the Tropicana brand name and turned it from a horizontal to a vertical alignment, making it more difficult to read, while at the same time boosting the size of the generic but now more legible claim that the product was "100 Percent Orange Juice."

The impact of the redesign was felt immediately. Less than two months after the introduction, with volume sales down 20 percent in a flat market, PepsiCo announced it would scrap the new design and return to the old one.[11] Unfortunately, the debacle occurred at the worst time. The economy was in the depths of recession, a time when private-label brands tend to gain share and consumers tend to hanker after the reassurance of nostalgic and familiar brands. Over the course of 2009, Tropicana lost share to Coca-Cola brands and private labels, with value sales dropping 11 percent.[12]

In the same *Advertising Age* article that reveals the sales decline, Andy Horrow, chief marketing officer at Tropicana at the time, is quoted as saying, "The biggest mistake we made last year was underestimating the passion consumers have for our business."[13] I doubt that passion had much to do with it. Far more likely, the majority of consumers, for whom Tropicana was the default purchase, felt some mild irritation on failing to see the familiar package where they expected it. The majority figured out what had happened and, at most, felt a mild pang of regret for the missing icon. Others either picked up another brand by mistake or, forced to reconsider their usual choice, deliberately chose a comparable brand, like Coca-Cola's Simply, or the cheaper store brand.

When your brand is reliant on its salience for competitive advantage, the last thing you want to do is change its design so dramatically that existing users cannot recognize it easily. On a final note, early in 2011, Tropicana swapped its traditional carton for a clear plastic container in the United States. This time, learning from its earlier mistake, the brand kept its logo, the orange with a straw, and its color scheme to provide continuity.

VISIBILITY AS AN ASSOCIATION TRIGGER

Particularly with habitual purchases like orange juice, simply being easily recognized is important. Modern supermarket and drugstore shelves are full of different pack sizes, brand extensions, branded competitors, and own-label lookalikes. At a bare minimum, a brand needs to stand out so that regular purchasers can find it when shopping on autopilot. But there is a far more important dimension to visibility, and that is a brand's ability to trigger positive and motivating associations seeded in people's minds by other elements of the marketing mix.

When consumers enter a store as shoppers, they do not suddenly become blank slates. They enter the store with ideas about different brands that are based on associations built up over time from advertising messages, word of mouth, and personal experience. They rely on these associations to simplify their decision-making. On average, Millward Brown has found that across a wide range of categories in a wide variety of countries, around two-thirds of people know what brand they want to buy before they go into the store, and around three-quarters of these "intenders" follow through on their intention. For shopper marketing to be effective, it needs to work with the predispositions people bring with them to the store, so a strong visual connection between customer's experiences with the brand both in day-to-day life and when in-store is important.

Is visibility important for brands that are not sold in a store? Walk down your local high street or drive past the strip mall, and I think the answer will become obvious: yes, it is. Unless a brand is quickly and easily recognizable and triggers positive associations, why is someone going to stop and check it out? You might argue that airlines do not need to worry about visibility—after all, if you booked online, what does it matter what the plane looks like? However, in 2001, when Rod Eddington, BA's chief executive, abandoned a $90 million project to repaint the airline's plane tailfins with "world images," he did so in order to "reinforce its Britishness."[14] Eddington clearly recognized that the stylized Union Jack that now decorates BA tailfins was far more meaningful to passengers than the disparate set of images designed to convey the scope of the airline's routes. In an interview with the United Kingdom's *Telegraph* newspaper, a senior BA executive said: "Rod feels that Britishness is at the core of this airline's appeal. We are a global carrier, but we are British and proud of it—and it is not just Britons who like our Britishness. Rod wants BA to be associated with Britain in the same way that BMW is associated with Germany, symbolising quality in a way that is understood worldwide."[15]

The need to trigger positive associations applies not just in the physical domain but also in the digital one, where if anything, it is even more important. In the always-on, fast-paced online world, where people decide almost instantly whether something is worth their time and attention, triggering an immediate positive response is mission-critical. Judgments are made fast and instinctively online. If people don't like the look of your website or don't recognize your brand name in search results or in a display ad, then the opportunity is gone in less than a second.

◼◼◼ SIDESTEP THE COMPETITION

In today's stressful and tiring world, the demand for an energy boost seems to be growing. Sales of traditional energy drinks like Red Bull and Monster Energy grew by 20 percent in the United States between 2011 and 2012.[16] Another potable source of energy has also grown in popularity over the last few years: energy shots. As the name implies, these products deliver the same energy boost as traditional energy drinks, but with less volume and without the "sugar crash." The category was pioneered and is now dominated by 5-Hour Energy from Living Essentials LLC. According to SymphonyIRI data, the brand makes up a combined 91 percent of the shot market, making it the third-largest energy brand behind Red Bull and Monster.[17] 5-Hour Energy's success does not just stem from creating a new energy drink format. Its unique approach to making the brand easily findable has also been critical to its success.

5-Hour Energy's two-ounce shots were first introduced in 2004 to General Nutrition Centers (GNC), a U.S. chain selling health and nutrition-related products. As described by Barry Silverstein for BrandChannel, once the brand proved popular with GNC customers, Living Essentials implemented a radical approach to extend its availability to more traditional retailers.[18] Recognizing that they could not compete head to head with brands like Red Bull, 5-Hour Energy sidestepped the competition and positioned the brand for sale by the checkout in convenience stores and garages. This strategy gave the brand high visibility, with the red bottles standing out from the rest of the odds and ends usually found on the countertop. The brand is now distributed in major retailers like Walgreens, Rite Aid, and Walmart and in vending machines—anywhere people might want an energy boost.

In-store visibility alone is not enough to guarantee success. In a rare interview with *Forbes* magazine, Manoj Bhargava, the creator of 5-Hour Energy and the CEO of Living Essentials, noted, "Getting (5-Hour Energy) next to the register isn't hard. Keeping it there? Very hard. Anything that sells, the stores will try. To own that space is really hard."[19] That the product lives up to its promise is part of the brand's success, but so too is heavyweight marketing. Targeting adults like truck drivers, security professionals, college students, athletes, and health-care employees, Living Essentials backs up its unique retail strategy with a wide variety of marketing communications, including TV and online spots, and sampling at clubs, bars, and liquor stores.

MAXIMIZE YOUR FINDABILITY

5-Hour Energy's success is a classic example of thinking through how best to maximize the findability of a brand. It is not just a matter of getting into stores, although that is the first priority for any marketer. It is also about making your brand easy to find and visible—so obvious that people do not even consider the alternatives. It is about making sure there is a clear synergy between marketing outside the store and what people see inside, since what the brand looks like and how well that image is imprinted in people's minds links the outside impressions to the point of purchase decision. Finally, you need to think about all the times and places where people might have a need for your product and make sure the brand is found there, instead of just aiming for 100 percent in your category's normal retail channels.

So ask yourself, is your brand everywhere it needs to be? What are the opportunities that lie outside the domain of traditional distribution channels? What are the new, growing retail channels that your brand might exploit? And what can you do to sidestep the typical purchase process to make your brand the only choice and not one of many?

Questions to Consider

1. Map out every moment when a consumer might have a need that your brand could fulfill or when a person might want to purchase the brand. Is your brand available at every one of those moments? Is it easy for the customer to find, and will it stand out in a competitive environment at that point? Is the brand instantly recognizable? How many clicks does it take to find and buy your brand online? Is that easy to do without encountering competitive brands along the way? Is your brand mobile-ready?
2. What opportunities can you identify to exploit the potential of the brand and make it easier for people to buy at the appropriate moments?

Chapter 11

Credibility

Two proverbs sum up the lessons outlined in this chapter. The first states, "Actions speak louder than words," and the second, "A man is known by the company he keeps."

Once you have identified what makes your brand's experience meaningfully different for consumers, then you should aim to align all the brand's actions to build and enhance that experience. This alignment develops your brand's credibility. Rest assured, if you are offering something meaningfully different now, your competition will either try to blunt that advantage or go one better. You cannot rest on your laurels. Instead, the brand should be refreshed and enhanced by innovation that is focused on that meaningful difference, and the launch of new product lines should benefit from, and feed back into, that difference.

Innovation cannot be limited to product. Rather, innovation needs to encompass every aspect of the brand experience—anything that changes the way people experience and interact with the brand. If your brand is often found in association with other organizations, brands, or people, then ideally they should share the same purpose or meaning as your brand. The association should boost your brand's credibility. In Chapter 9, I referenced the ice cream brand Ben & Jerry's. The brand's heritage has always meant that it takes a stand on social issues, so when the board of Ben & Jerry's announced that it stood with the Occupy Wall Street protesters in 2011, the statement was entirely in line with the company's ethos and created some nice publicity.

RADICAL VERSUS INCREMENTAL INNOVATION

Although one should never stop looking for the next game changer, the truth is that such discoveries do not come along very often. If they did, then most of the product categories in this book would be in a constant state of flux instead of a constant state of stalemate. As noted before, what tends to happen is that a game changer shakes things up and then things revert to a more stable state.

Even the categories subject to the fastest change, like those related to technology and the Web, follow the same pattern. The successful disruptor, be it Amazon, Netflix, or Apple, holds onto its dominant position until the next game changer comes along, but that does not mean that innovation ceases. Amazon, Netflix, and Apple continue to innovate, but the innovation is incremental, not radical, and it tends to play to the brand's strengths. Amazon is seeking to make same-day delivery a reality. Netflix sidelined Blockbuster with its shift to streaming movies. And Apple continues to refine the iPhone, making it even easier to use with Siri voice control. This type of innovation is often referred to by the Japanese word *kaizen*, which implies a continuous dedication to improvement across all aspects of a business.

In a post on the *Harvard Business Review* blog, Patrick Vlaskovits not only casts doubt on whether Henry Ford actually uttered the famous line "If I had asked people what they wanted, they would have said faster horses" but also details how Ford's failure to innovate led to the loss of the company's dominant position in the U.S. automotive market.[1] Vlaskovits notes that Ford's success originated from the application of mass production to cars, but that in doing so, Ford essentially froze the design of the Model T. Vlaskovits contends that as long as the Model T remained ahead of the competition, Ford's was a winning strategy that allowed the company to cut costs and drive growth. But Ford continued to pursue the same strategy when the competitive context changed. The advent of General Motors, which produced new cars for distinct market segments and made annual changes in design, smashed Ford's hold on the market. According to Vlaskovits, in 1921, the Ford Motor Company sold about two-thirds of all the cars built in the United States, but by 1927, that share had fallen to about 15 percent, following an attempt to retool and respond to these innovations.[2]

Just as it is fatal for a brand to ignore the changing nature of competition and technology, so too is it important that a successful brand play to its strengths and seek to improve the experience that made it successful—not change it beyond recognition. If the fundamental need has not changed, then evolution, not revolution, is the name of the game.

BUILDING CREDIBILITY THROUGH LUXURY

In Chapter 9, I reviewed the success of Häagen-Dazs in building a strong brand in China. Now I would like to revisit that brand in the context of credibility. I think the Häagen-Dazs case study proves that once you have identified what your brand stands for, the decisions related to how to promote it become far easier to make. Each of the actions outlined in the following paragraphs serves to reinforce the brand's credibility as a luxurious treat to be shared with loved ones.

Häagen-Dazs has over 160 ice cream stores in China, but as a luxury brand, the location of each one had to be carefully selected. Would Natalie shop there? Walk along the Nanjing Road Pedestrian Street in Shanghai, and you will find a store located under Le Royal Meridien hotel. In Beijing, there is a Häagen-Dazs right next to the Apple store, on the corner of the famous Wangfujing Street. These locations are expensive and deliberately chosen to associate the brand with other foreign luxury brands. Häagen-Dazs walks a fine line between expanding to meet demand and expanding to be in the right places.

Just as the store location confers an image of fashion and luxury, so do Häagen-Dazs Wedding Cakes. "Celebrate your love in an unforgettable way. That's the Häagen-Dazs promise," states the brochure for these concoctions of cake and ice cream. The "Royal" cake is up to eight layers tall and promises to add "a touch of extravagance to your wedding." The "Rainbow Couture" cake is decorated with colored chocolate and claims to express "your true love's colors." And the "Dream Castle" cake will be "your Castle of Joy on which to pledge your everlasting love." Costing up to the equivalent of nearly US$5,000, these cakes are delivered in a special van and served by uniformed Häagen-Dazs staff. This is not a business line that Häagen-Dazs necessarily wants to expand, but it provides an important showcase and associates the brand even more closely with romance and celebrity status.

Unlike many multinational brands seeking to be successful in China, Häagen-Dazs has eschewed the tactics of smaller pack sizes and cheaper formulations in favor of doing everything it can to justify its price premium. Higher prices connote quality—particularly in status-conscious China—but the price of even the simpler desserts on the store menus is expensive by anyone's standards. That, however, makes them all the more significant as something to share with a loved one or give as a gift. For Häagen-Dazs, the question is not how to make the brand more affordable, but how to use its price to help build credibility as a luxurious treat.

Moon cakes are a traditional Chinese baked good eaten during the Mid-Autumn Festival. They are a popular festival gift in China, and today, ice cream moon cakes account for 30 percent of Häagen-Dazs's sales, even though the festival only lasts a week each year. Once again, the Häagen-Dazs team has adroitly tapped Chinese culture to provide a meaningfully different brand experience, one that builds on what the brand stands for and extends its footprint outside the store.

■■■ BUILDING CREDIBILITY THROUGH INNOVATION

Although it is in a very different business and serves very different customers, IBM uses innovation to build its credibility as a leading-edge technology company. Watson is a cognitive computer system designed by IBM to solve one of today's biggest problems: data overload. The amount of information available to us is growing at an incredible rate, and most of it is in the form of unstructured data text and spoken words, not coded data. Creating a system that can search, parse, and develop answers from unstructured data could have enormous benefits for many industries, particularly financial services and health care. But how best to promote such a system? What would capture the popular imagination and make Watson famous?

As explained on the IBM website, the popular game show *Jeopardy!* was selected as the ultimate test of the machine's capabilities because it relied on many human cognitive abilities traditionally seen as beyond the capability of computers.[3] Winning the game would be an important demonstration of the system's potential and offered an opportunity to enhance IBM's reputation. Of course, the show's obvious popularity was another positive, but there was also a downside. Beating the past winners of *Jeopardy!* might also evoke negative responses. People might view Watson as a potential job killer or, thanks to its sci-fi archetypes, a humanity killer. IBM and its agencies were going to have to handle the publicity around the event very carefully.

Popular though *Jeopardy!* was, IBM still needed to publicize the event in order to achieve the maximum return on its investment. As detailed in the 2012 Gold Effie case study "Watson," IBM's communications team and advertising and marketing agency Ogilvy & Mather needed to ensure that the three-night special was viewed and would generate positive buzz in both traditional and social media.[4] While the ultimate objective was to commercialize a new marketplace for natural language processing technology, in the short term, there was money to be made by selling the underlying hardware, analytics software, and consultancy that powered Watson's unique capabilities.

To make the most of the event, IBM needed to make the science meaningful and portray humanity, not the machine, as the winner. The project consisted of an integrated three-month campaign that told the Watson story on IBM's terms. Material collected over the prior two years of the project was used to create 22 videos that portrayed the human story behind the project.

These videos were used to stimulate interest and start a public conversation before the event. A visual identity derived from IBM's Smarter Planet symbol was created for Watson and animated to represent the computer's confidence in its answers in order to engage the audience during the event. This made Watson more appealing without making him seem creepy. A multimedia ad campaign publicized the show itself, and in the weeks after the show, print and online campaigns explained the future applications of the technology while sales materials were developed to focus on both future and current technology.

While the research and development budget behind Watson was immense, the paid media budget was estimated at between $5 and $10 million, a fraction of the company's overall marketing spend. However, the investment paid off by generating over $50 million in additional earned media value. More important, the Effie Case Study reports that Watson drove leads and potential lead revenue into IBM's demand system, resulting in a profit payback of three times the investment.[5] These numbers only scratch the surface of Watson's full potential, however, since Watson is only now being deployed in the health-care and financial services industries. Watson's success on *Jeopardy!* in 2011 was a massive step forward for computing, but it represented more than a test of the new system. Participating in the popular game show helped publicize and enhance the credibility of the system and IBM in a way that no other communications vehicle could have done.

▄▄▄▄ BUILDING CREDIBILITY THROUGH ASSOCIATION

While by no means as fundamental to success as innovation, association is another important mechanism by which brands can build credibility around their meaningfully different experience. In some cases, the association may have a direct business objective, as when the Mexican value airline Volaris partnered with Southwest airlines to establish U.S. routes. But there is another important dimension to any partnership: how well it fits with what the brand stands for in people's minds. Southwest is the best-established American budget airline, and it represents a good fit for Volaris in terms of positioning and character.

Some associations have a more indirect influence on the business by helping to define what the brand stands for. The fashion brand Burberry has long been associated with celebrities like Kate Moss, Emma Watson, and Rosie Huntington-Whiteley. True to this strategy, the new Burberry spring/summer 2013 campaign features several British models, including

Edie Campbell and Charlie France. But the campaign also features Romeo Beckham. The ten-year-old son of Posh and Becks is an integral part of the campaign, lending it a youthful enthusiasm that serves as a counterpoint to the more poised and posed models. The inclusion of the young Beckham not only brings important energy to the campaign but also draws on the allure of his parents and indirectly highlights the brand's children's wear line.

◼◼◼◼◼ VOLARIS: POSITIVE ASSOCIATIONS IN FLIGHT

The last time I was in Mexico City, I had the chance to sit down with José Calderoni, marketing and sales manager for Volaris, and he explained a little about the airline and its approach to business and marketing. Based on what he told me, I would suggest that Volaris is a great example of a value brand that recognizes that cheap does not have to mean bad. But it was the fact that the brand's publicity was focused on its core brand experience that really caught my attention. When flying is your business, why not make use of your planes in your publicity events? Calderoni mentioned a fashion show to promote new uniforms that was held onboard one of the company's planes and then continued:

> But the craziest thing we have done is two live concerts onboard our aircraft. The first one was done in conjunction with Coca-Cola, and it was two bands playing one after the other. The second time we did it was to celebrate the flight to Los Angeles in the USA, where there is a very large Latin American community. We asked ourselves which is the band that best personifies the relationship between the U.S. and Mexico. And there is this one band called Los Tigres del Norte, and their songs often feature stories of people who have migrated across the border. So we thought, this is perfect. We had them onboard for about an hour because they said, "Either we play for an hour or we don't play." So we said, "OK, we'll play an hour." We had to divert the flight a little to make sure that by the time the plane got to U.S. airspace the concert was over.[6]

Needless to say, this sort of event creates a great deal of publicity for the airline, but the choice of Los Tigres del Norte lent the event a meaning beyond simply creating buzz. The choice tapped into a well-known cultural phenomenon and created a sense of empathy with the target audience. Also note the reference to Coca-Cola as a co-sponsor of the first concert. Volaris has a marketing agreement with both the Coca-Cola Corporation and Cuervo. These two brands essentially sponsor the in-flight service, allowing Volaris to serve premium brands that they might not otherwise offer and enabling the partner companies to try out new products and promotions onboard. As Calderoni notes, such sponsorships are an innovative way to bring premium brands onto Volaris flights to enhance the passengers' experience at the minimum possible cost.

◼◼◼ DIMENSIONS OF ASSOCIATION

In the ValueDrivers workshop, we identify seven common ways that a brand might benefit from association: celebrity, an environmental issue, charity, a cultural event, a social issue, a sports event, or another brand. We ask the participants to identify examples of each association that fit with their brand's meaningfully different experience. Providing an example for each type of association initially seems a daunting task, but it usually proves easier than expected and produces remarkable consistency between groups once that people have clarity regarding what their brand stands for.

That said, seeking to grow a brand by association is not without challenges. As more brands seek to engage their consumers through the power of association, they would do well to bear in mind the three F's: fit, focus, and fame.

Fit

Whether it is celebrities, events, or branded content, the stronger the perceived fit between the brand and its partner, the more positive the impression of the brand will be.

In an analysis entitled "Celebrity Power: Can Less Be More?" my colleague Dede Fitch notes that today, people don't have a high degree of trust in major celebrities.[7] When she wrote the piece in 2007, Lance Armstrong had just received the top endorsement in the NPD Group's Celebrity Influence Survey as "someone I trust," even though he was endorsed by fewer than one in five people. Five years later, I daresay the discredited Armstrong is not topping the poll. Dede notes that in contrast to big-name stars who carry big-name risk, it was the lesser-known celebrities who were cited most often as having a positive influence on a purchase decision.

Dede proposes that new niche celebrities can better act as spokespersons to which ordinary people relate because they lend authenticity to the brand's communications. Mike Rowe, the host of the Discovery Channel's *Dirty Jobs* show, is a good example of this. In each program, Rowe explores a dirty job and attempts to complete the same task as the people whose job it really is. Rowe is associated with Wolverine World Wide's Cat Footwear, the global footwear licensee of Caterpillar® Inc. On the Cat website, a page dedicated to Mike Rowe announces that if Rowe ever writes a book on lessons learned from people with dirty jobs, chapter one will be "Don't skimp on your boots."[8] The association makes perfect sense for Cat.

If ordinary people can relate better to niche celebrities than they can to megastars, then there is one thing to which they can relate even better: ordinary people doing extraordinary things. In her analysis, Dede cites the example of fast food giant Subway, which has leveraged such authenticity to great effect with their U.S. spokesperson Jared Fogle. Now also known as the Subway Guy, Fogle lost 245 pounds by adhering to a diet of his own creation: two low-fat Subway sandwiches every day. After using Fogle's story in their advertising, Subway was deluged with testimonials from people who were inspired by Fogle to lose weight. Fogle communicated not only a differentiating benefit for Subway—healthy sandwich options—but also a message of empowerment.

Focus

Brands have been seeking to benefit from the association with movies and television shows for decades. However, as advertisers struggle to engage the attention of today's overmarketed consumers, they are looking with renewed interest at the placement of brands in various entertainment genres. Many options are available, ranging from simple product placement to brand integration, branded entertainment, and, ultimately, branded media channels like those on YouTube. Each option has its own requirements in terms of planning and investment, and each offers different risks and benefits for brands. There is a vast difference between paying a product placement firm to represent a brand to a film's property masters and negotiating a brand's integration into the storyline.

That is why many brands are creating stories that feature their brand and then having them made into programs. Take the example of Dove in China. Finding that the brand's Campaign for Real Beauty was too dependent on Western attitudes and tensions for Chinese consumers and faced with a cluttered media environment in China, Dove took a new approach, launching a Chinese version of *Ugly Betty*, a show known in America from the ABC television comedy of the same name but originating from a Colombian *telenovela* (which in turn found its origins in the Hans Christian Andersen fairy tale "The Ugly Duckling"). The story of Ugly Wudi, who uses her brain to triumph over male chauvinism and company politics and win the heart of her boss, is built around the Dove brand. Wudi works at an ad agency charged with creating a campaign for Dove, a setting that provides many opportunities to feature the brand. The program series proved highly successful, attracting large audiences.[9]

Given the need for multiple partners to bring a successful integration to fruition, marketers must maintain a relentless focus on the project's communications objectives. This does not, however, mean that the strategy should be apparent to the intended audience. Gratuitously highlighting the brand's benefits in the same way as Ugly Wudi does will often work to the detriment of both brand and content appreciation, particularly in cultures like those of Western Europe, which are less accepting of advertising and brand promotion. Instead, brands need to seek opportunities for active integration into the plot, with the action focusing attention on the brand and providing implicit communication of the brand's desired positioning and benefits.

Fame

The problem faced by most forms of association is that the footprint of an individual association is often quite small. Heineken is reported to have paid over £28 million (or nearly US$45 million) for James Bond to take a sip of their beer early on in *Skyfall*, and yet, even though I knew of the placement ahead of time, the moment passed unnoticed by me when I watched the movie. In retrospect, the only reason I knew Bond was drinking Heineken was that I had read about it in the trade press.

If a brand wants to extend its reach beyond the aficionado, it may need to use other media to make the event famous, just as IBM did with Watson's participation in *Jeopardy!* When well managed as part of a broad marketing program, the reach of sponsorship can be extended beyond the passionate few who are inclined to actively engage with a brand to the vast mass of buyers who are willing to spend money on it. Marketers need to seek ways to amplify and disseminate successful integration efforts via pass along, word of mouth, and attendant advertising and publicity.

◼◼ BUILD CREDIBILITY WISELY

Innovation is a powerful mechanism by which to reinforce a brand's meaningful differentiation. Rather than simply seeking "new news," marketers need to focus on creating news that enhances what the brand stands for. The same is true of association. Brands need to choose their partners wisely. They do not only need to enhance the brand's reputation now—thought must also be given to the future. What is the risk that a big-name star will prove fallible? And even if the relationship continues to be beneficial, might it become limiting? Subway's relationship with Jared Fogle came about more by accident than by design, but he has starred in over 50 TV commercials over the span of a decade with a positive effect on sales.[10] However, Fogle's strong association with dieting presents Subway with a dilemma. It is tough to use Fogle to advertise the brand's five-foot subs.

Questions to Consider

1. What does your innovation pipeline look like? How do all the potential innovations add to or reinforce the meaningfully different experience your brand offers? How could they be shaped to do this? What relationships does the brand have with people, programs, and organizations? Does each relationship reinforce the meaningful difference of the brand?
2. Can you identify new innovations and associations that would directly support the meaningful difference?

Chapter 12

Vitality

A successful brand must have vitality; it must seem active and alive, part of the contemporary scene. A brand's communications must keep the brand salient and current if the brand is to remain top-of-mind for consumers. One of the most important qualities that people look for in a brand is popularity, even this popularity only extends to their own clique. The use of social media can help a brand seem contemporary and encourage people to talk about it, as long as the brand creates experiences with inherent talkability. Communities, offline as well as online, are powerful sources of human connection; however, the choice of communication channels and creative content should be based on and driven by the brand's meaningful difference. Vitality is also served by maintaining a fresh look and feel to packaging, logo, and communications, though innovation in these areas needs to be balanced by the need to maintain clarity and identity.

There are many myths about what makes "advertising" successful and, in particular, about what makes digital advertising successful. The essential point is summed up in the following quote from a 2002 *Advertising Age* special report that lauded JetBlue as Marketer of the Year: "Advertising is the last thing you bring to the mix," says Amy Curtis-McIntyre, vice president of marketing at JetBlue. "You start by getting the product right, getting your attitude right, getting everyone internally understanding the mission. Then you move to telling the story through PR. You build the advertising last."[1]

Great "advertising" is founded in an understanding of both the brand's meaningful difference and its empathy with its target audience. The advertising originates from the brand; it does not create the brand.

██████ CHANGING THE ODDS, NOT PLAYING THE ODDS

In Chapter 4 we examined the evidence that brands that punch more than their weight—that is, have a share of media voice higher than their share of market—are more likely to grow market share in the following year than those that do not. Several different studies come to a similar conclusion, which suggests the empirical generalization that overinvesting in media one year will pay off the next year at a predictable rate depending on category and brand size. Peter Field sums up this idea as follows: "The lesson for account planners is clear: brand targets as well as campaign and agency assessments cannot be divorced from the communications budget. A target that is realistic with one budget level can become unrealistic when this is lowered. And a campaign that might have generated strong growth with sufficient budget can fail without it."[2]

Field acknowledges that this conclusion runs counter to what most marketers are hearing from their senior management: do more with less. All the evidence suggests that, on average, what is required to do more is *more*. The odds are that if you lower your SOV below your SOM, then your brand's market share will decline in the following year.

That may be a depressing finding for many marketers, but there is one way out of the budget trap. Don't just spend more; spend more efficiently. For a given spend, the variable that has the biggest influence on marketplace outcomes is copy quality—how well exposure to the advertising conveys a salient and positive impression of the brand. So rather than playing the odds, marketers should seek to change the odds in their favor by creating impactful and engaging advertising. This may be easier said than done, but it is probably a lot easier than getting that 10 percent increase in marketing budget out of your CFO.

TELL YOUR BRAND'S STORY

What is the best way to advertise a brand? Well, if your brand is meaningfully different, then let its story speak for itself. Understand the difference that your brand makes in people's lives and dramatize it. By dramatize, I do not mean exaggerate. I mean make it apparent. Don't make people think—help them feel and anticipate the brand experience. Make it memorable. And keep it authentic, true to the brand.

This is what Andrew Stanton, writer and director of *Finding Nemo* and *WALL-E*, said in a TED Talk entitled "The Clues to a Great Story."[3] After demonstrating the link between a story from his own life and a key scene in *Finding Nemo*, he said: "And that is the first story lesson I ever learned: use what you know. Draw from it. It doesn't always mean plot or fact. It means capturing a truth from your experience, expressing values you personally feel deep down in your core."[4]

That is what advertising needs to do. It needs to capture the truth and values of the brand. Some of the most successful advertising in the world succeeds because it focuses our attention on what matters most: a positive experience delivered by a product making good on its promise.

FUNCTION AND EMOTION

Take the iconic ads for the iPod that show silhouettes of people dancing to the device's music. Nothing is said about how the product functions, yet the ad ably conveys the experience offered by the brand. It reminds me that there is a reason why we talk about ad "impressions." The ad does not tell you how easy the iPod is to use or how many gigabytes it stores. You just get an impression of how much fun the experience of owning one could be. It conveys both the product's primary function—portable music—and the emotion evoked by using it.

In the last few years, agencies have aggressively pursued an agenda to promote the need for emotion in advertising. In the process, they seem to have driven a wedge between function and emotion in our minds, and by doing so, they may have created a trap that dooms much advertising to be less effective than it might be otherwise. Functionality has become equated with rational arguments and a litany of benefits, specifications, and capabilities. Emotion has become associated with implicit communication and unconscious "persuasion." Yet the most powerful ads, like the one for the iPod, recognize no such distinction, because they focus on conveying an impression of the brand experience.

No one can deny that a relevant and emotionally charged ad is going to be more effective than one that, no matter how relevant, fails to elicit an emotional response. All too often, however, emotion is enlisted to engage the viewer with the creative vehicle, not with the brand. Yes, we need people to engage with the ad, but just as importantly, we need to convey a compelling impression of the brand. If advertising is to be effective, then people need to understand or be reminded of how the product makes a difference in their lives.

Making the viewer "live" the experience in some way is critical to a brand's success. The more emotionally resonant and personally relevant the portrayal, the more effective the ad will be. An ad should address the following questions: How does it feel to use the product? What are the sensory experiences that one could expect from using it? What are the rewards to be gained from using it? Remember, any ad is speaking to two audiences: users and potential users. On seeing an ad, users should feel a sense of recognition—"That's how it feels." They should identify with what is shown or said. Potential users should feel a sense of anticipation and interest—"Is that how I might feel?"

BEYOND ATTENTION

In his book *The Feeling of What Happens*, the eminent cognitive scientist Antonio Damasio makes the point that for feelings to have subsequent influence, one must not only experience feelings but also be conscious of them at the time they are experienced.[5] In other words, just because we attend to something once does not mean that we are going to remember anything about it at a later date. Only when facts, ideas, and impressions are emotionally charged will they create a lasting memory. The stronger the emotional charge, the more likely we are to reflect consciously on the experience and the more memorable the event will be.

Does an ad need to make people feel what makes the brand meaningfully different? No. Many brands get by simply by being salient. But the ones about which people care, the ones on which people are willing to spend more money, are the ones that exemplify what their brand stands for in their advertising.

ENGAGE; DON'T PREACH

All stories have a purpose. At the very least, the purpose is to entertain. The best stories, however, are enthralling, inspiring, and even educational. They add value to our lives. We want to share them and talk about them. They are meaningful to us. However, unlike many TV commercials, stories do not need to explain what they mean. Instead, they involve us. They give us what we need to make sense of the story and own it. Our interpretation makes it ours.

Andrew Stanton suggests that storytelling without dialogue is the most inclusive approach that you can take. He says, "We all want to work for our meal when we watch a movie; we just don't want to know that we're doing it."[6] Stanton goes onto say that humans are compelled to deduce and deduct, adding that "it is the well-organized absence of information that draws us in."[7]

All too often, I think that advertising has become the domain of the preacher, not the storyteller. In video ads, the voiceover is a means to make up for the fact that the story is badly told. Print ads heavy with copy are struggling to identify what really matters. These are the client safety nets that promise a result even if the audience does not understand the basic idea. Unfortunately, all the evidence suggests that if people don't easily understand an ad, then it does not matter what is said—they won't remember it.

BOGS ACROSS AMERICA

There are not many marketing campaigns in which the central theme is a sea of red berries, but that image is exactly what lies at the heart of Ocean Spray's "Straight from the Bog" campaign. This marketing campaign is a great example of a brand using multiple touch points to focus the public's attention on its meaningful differentiation: cranberries look good, taste good, and are good for you.

For those not familiar with these tart, red berries, let me just take a moment to explain that cranberries are a big business in North America. The berries grow on low-lying bushes. When the time comes to harvest them, the growing beds are flooded until the bushes are underwater. A water-reel harvester is used to knock the berries from the bushes. The floating berries can then easily be corralled and collected. The harvest makes a striking scene, with acres of red berries floating on the flooded growing beds.

Ocean Spray is an agricultural cooperative owned by more than 600 cranberry growers. It is North America's leading producer of canned and bottled juices and juice drinks, and it has been the bestselling brand name in the canned and bottled juice category since 1981. In 2005, Ocean Spray shared a common problem with other manufacturers of fruit juice. The American consumer's belief in the health properties of fruit juice had been undermined by negative publicity and a craze for low-carb diets. Juice sales had declined 20 percent between 2000 and 2005, and juice marketers had responded with the introduction of diet variants, flavored waters, and vitamin drinks, as well as new beverages claiming "superfood" status. Ocean Spray's sales were down more than 22 percent; furthermore, unlike other companies, Ocean Spray's imperative as a cooperative was to sell cranberries, not diversify into other beverage categories. The brand faced a big challenge: How do you revitalize interest in cranberry juice?

As detailed in the 2008 Gold Effie–winning case study "Straight from the Bog," Ocean Spray needed to counter the health claims of other beverages with one of their own.[8] This was uncharted territory for the brand, since in the past, taste and refreshment had proved a more compelling story than health. Besides, the major health benefit associated with cranberry juice—fighting urinary tract infections—was hardly attractive to consumers.

Luckily, however, the cranberry comes loaded with antioxidants and vitamin C, and the only real question was how best to position this message. Scientific and consumer research led to a health claim that Ocean Spray could own: "Powerful nutrients that cleanse and purify your body and help strengthen your immune system." ''

With that frame in mind, Ocean Spray's agency, Arnold Worldwide, then needed to figure out how to communicate the message's story in a way that would resonate with consumers and remain true to the 75-year-old Ocean Spray co-op. To do so, the agency created Justin and Henry, two cranberry growers who are always seen wearing waders and standing thigh-deep in floating cranberries. Henry is the straight man of the duo, delivering the message, while Justin gives that message a wacky spin. For instance, in one 2010 TV spot, Henry delivers the message that Ocean Spray cranberry juice is rich in powerful nutrients, and then Justin announces, "Cranberries are the ninja fruit" and leaps into the air in an imitation karate move. The commercials, although simple and message-centric, proved to have a strong branded impact, making efficient use of the media spend.

An important and unique element of the campaign was an event-marketing program called "Bogs Across America." Cranberry "bogs" were set up in major city centers like Rockefeller Center in New York City, Pioneer Court in Chicago, and Hollywood & Highland Center in Los Angeles. These bogs consisted of 1,500 square foot freestanding ponds containing more than 2,000 pounds of cranberries and staffed by growers who educated consumers on the health benefits, versatility, and heritage of the cranberries. The sight of the floating cranberries created a visual link between the TV campaign and the event and provided a focal point for product sampling and cooking demonstrations. The events helped create strong publicity and talkability around the campaign and generated 1.3 billion media impressions in two months.

The "Straight from the Bog" campaign reversed equity and sales declines. The impression that Ocean Spray helps cleanse and purify your body increased significantly, and perceptions of quality and trust also improved. The Effie case study reports that base volume measured by IRI (Information Resources, Inc.) increased 11.6 percent in 2006 versus the prior year, while overall category volume decreased 5.6 percent during the same time frame.[9] Testimony to the long-lasting success of the campaign is the fact that it is still going strong in 2012, with "Bogs Across America" having become an annual event.

In many respects, the initial "Straight from the Bog" campaign was a pretty traditional one. It led with TV and was backed up by radio, print, and transit ads to generate reach. The focus on events and sampling added an experiential element that was high impact, if low reach. Merchandising ensured in-store visibility. But what of the digital sphere? The Effie case study simply lists "website" under its "Interactive" category. Five years later, no brand would consider launching a campaign without digital, and for good reason. But are we using digital as effectively as we might?

DIGITAL'S POWER

Over 2 billion people worldwide now have access to the Internet, or about one-third of the global population. Seismic shifts in technology and communications are affecting the future of traditional media like newspapers and magazines in developed Internet markets. Network TV has yet to suffer in the same way but will surely be affected in the near future, as online is now challenging TV for the top media slot in the United States. In 2012, Google alone generated more ad revenue than all newspapers combined and nearly as much as magazines, and things are changing within the digital domain. Social networks are today's place to spend time. Facebook reaches one-seventh of the world's population. In a few years' time, smartphones will be the dominant means by which people surf the Web and interact with apps and ads.

We know from Millward Brown's own databases how powerful paid-for advertising of all kinds and across all digital platforms can be in generating stronger brand associations, consideration, and purchase intent. We can easily see the power of social media in creating spaces for fans to gather and fueling their ability to spread the word. We have experienced the creative potential of video to generate powerful viral effects. The Web also allows different forms of activity to be connected simply by several clicks—from an ad to a microsite to a competition to a purchase.

Just like any successful new product, the Web has been powerfully sold to potential online advertisers through the numbers—the eyeballs, the clicks, the case studies, and other research data. It is often called the world's most measured medium.

DIGITAL'S PERIL

The digital domain presents some big challenges. For all the successes, there are many more failures. Most video does not go viral. Many ads are invisible. A surprisingly high proportion of ads have negative effects. Few fan pages garner many fans, and those that do are often simply collecting heads, not building brands. Few people, when offered the opportunity to click from one space to another, take that opportunity. Material created for one domain is unthinkingly transferred to another with negative consequences—ads and sites not optimized for mobile, for instance.

Measurability creates evidence, but it also creates accountability. We are moving from an era of hype to an era of balance. Nothing in the digital space is a magic bullet—everything has to be worked for. That means we need to understand what works and what does not. More specifically, we can benefit from understanding how people behave online and how that behavior influences their response to communications. These days, there is no difference between digital and traditional; the digital world is no longer a separate domain with which marketers can experiment—it is an integral part of people's lives. With its rise, the power of traditional media channels like print and TV is waning, or at least morphing into a digital equivalent. We must become as adept at marketing in the digital domain as we are in the traditional media channels.

Right now we have the luxury of apparently lower production costs, more accurate targeting, and relatively cheap impressions in digital, but these advantages simply allow us to reach people in a cost-effective way; they do not guarantee an impact. Efficient return on investment depends as much on the content as on how well we reach people. If we are to generate a better return on our digital investment, then we must understand the motivations and instincts behind people's digital behavior and apply that learning to make our executions more effective.

◼◼◼◼ FAST AND INSTINCTIVE DECISION-MAKING

We use the phrase *"lean forward"* to describe our approach to the digital domain, because that phrase signals a greater degree of attentiveness and concentration on behalf of the user than might be found in traditional media.

In part, this reflects the physical need to attend to a smaller screen, but to a greater degree it reflects the fact that the user is in almost complete control of the content that he or she consumes. Unlike the sequential presentation of live TV and radio, digital allows users to choose what content they want to encounter and when. Users do not want to waste precious time, so they need to judge quickly whether content is relevant or not. Judgments are made fast and instinctively.

This leads to some very different dynamics when it comes to ad effectiveness. All our work with online and mobile advertising finds that the most effective ads meet three criteria.

1. Digital ads need to communicate their relevance very quickly

Our eye-tracking database suggests that, on average, only 50 percent of people even look at an online ad when presented in the context of a natural Web-surfing experience. People who see generally look at it for 2.7 seconds.

2. Digital ads need to be branded from beginning to end

Purchase intent is over 50 percent higher when the brand is featured throughout a display ad. Why? Because you cannot guarantee when someone will look at the ad. Many online advertisers fall into the trap of assuming a linear mindset and an attentive audience when creating ads, so they only reveal the brand toward the end of a display or video ad. In reality, most people have stopped looking by the time the brand is revealed.

3. The message needs to be simple and (ideally) visual

Pictures do save a thousand words. A quick glance at a picture will convey far more information than if someone is required to scan a series of words. Find the right visual to communicate your brand idea, and it will work for you even if someone only glances at the ad. Don't underestimate the importance of cues in the online arena. Visual cues can help the ad stand out from a crowded webpage, signal which brand is being advertised, and deliver the message.

The need to fulfill these three digital success criteria make it even more imperative to know what your brand stands for in people's minds. That is the foundation from which you need to build an impression using cues that will clearly signal the brand and its message.

MAKE YOUR FIRST IMPRESSION COUNT

The great thing about digital is that it provides for linkage and interaction. If someone is interested in your brand, then he or she can explore content and delve deeper. Of course, it is great when people do want to engage with your content at length, but recognize that these people are the small minority. Not everyone is going to play your online game, download your app, or make a video—those people number in the thousands at best. Plan for that level of engagement, but in addition, make sure that the millions of typical exposures are not wasted.

If people are making judgments about what to view so quickly, then the first second of exposure must be as impactful as possible. Marketers need to ensure a good first impression for everyone who has the opportunity to see the ad, not just the few who decide to engage with it at length or click through to other digital assets.

The truth is that few people are at all interested in thinking about brands when exposed to an online ad. Instead, they want to get the news, find recipes, check out information on celebrities, and connect with their friends. However, they are willing to engage with content that they find meaningful and different. Think about viral advertising. What is the motivation for people sharing content with others? They gain kudos if they share content that their friends judge to be relevant and engaging. Videos like Melbourne Metro's "Dumb Ways to Die," Pepsi Max's "Uncle Drew," or Nike's "My Time Is Now" become social objects shared and liked by hundreds of millions of people around the world.

A MEANINGFULLY DIFFERENT BRAND IS LIKED ON FACEBOOK

Posting fun and involving content is important, but these are table stakes if a brand wants to make the most of its social media presence. Millward Brown's BrandZ data show that social media fans spend nearly five times the amount on their favored brand than nonfans do. In all likelihood, that extra value predates these people becoming fans. They like or follow a brand because they find it to be meaningfully different in life, not just on Facebook. For instance, the number of fans that a U.S.-based airline will have on Facebook can be predicted with accuracy from how many people say they have a high opinion of the brand in BrandZ.

Once you have allowed for the overall propensity to like or follow a brand, our own FanIndex finds that two overall criteria separate a great social brand from an average one on Facebook. If they are not solely interested in freebies and giveaways, then what fans really want is content that is personally relevant on the one hand and provides a sense of being part of a larger community on the other. The former requires the brand to post content that is varied and meaningful to its fan base, while the latter requires the brand to encourage interaction between fans.

Burberry Uses Digital to Revitalize Its Brand

Burberry is often credited with having "democratized" fashion. In the past, luxury brands eschewed social media because the nature of the medium made it impossible to curate, control, and protect the brand as had been traditionally done. But the archetypal British brand broke the

➡

mold and gained an advantage by reaching out to fans through social media.

There was a time when Burberry stood for little more than its trench coat, but in recent decades, the signature Burberry check started to appear on a wide range of clothing. The check was distinctive and easily recognized and ensured that people knew which brand of clothing they were seeing, but this visibility also brought undesirable side effects. In the United Kingdom, the brand became associated with football hooligans and risked being brought downmarket. Association with latter-day "chavs," or soccer hooligans, further threatened to undermine the brand's exclusiveness, and with the onset of the Great Recession, people wanted a different aesthetic. They became less focused on the overt symbolism of status and more desiring of clothes that were elegant, cool, and relaxed.

Fortunately for the brand, Burberry could deliver on that need, thanks to business decisions made in the preceding decade. A decision to end all licensing of the brand, something that had been prevalent until the mid-1990s, helped Burberry regain control of its own destiny. Design decisions could now be reflected throughout the brand's product lines, without the need to negotiate with other companies. Today, the brand Burberry stands for "democratic luxury," and while the iconic trench coat remains ubiquitous, the check is far less prominent.

Celebrity endorsement, print advertising, and now social media have played an important role in the brand's comeback. Burberry was one of the first luxury brands to recognize that the time had come to turn social media from a threat to an advantage, but in doing so, the brand was simply recognizing a fact of life. Three quotes sum up the brand's strategy:

- "Attracting the Millennial customer to luxury started two years ago—I said that we can either get crushed or ride the greatest wave of our life."[10]
  —*Angela Ahrendts, Burberry's chief executive, March 2010*
- "We always said if we were going to target a millennial consumer then we had to do it in their mother tongue, which is digital."[11]
  — *Ahrendts, December 2012*
- "Digital is something that should never be an afterthought—it should never be a checklist that is something to do. Digital for me is not a project; digital is a way that we live. If you deal with it as a project, it will always be superficial."[12]
  —*Christopher Bailey, Burberry's chief creative officer, January 2013*

Digital has become an integral part of the Burberry brand experience, from "Art of the Trench," a campaign that integrates user-generated photos to document the classic trench coat, to the vast array of clothes carried on the company's website (reported to be ten times what might be found in the typical store).

The brand's commitment to social media has had measurable results. Typically, there is a good relationship between a brand's equity and the number of fans and followers it has. A comparison of BrandZ equity data with the number of likes a brand has on Facebook finds this association to be true of luxury brands as well, with one notable exception. At the end of 2012, Burberry had over 14 million "likes" on Facebook, more than any other brand—including Louis Vuitton and Christian Dior—and over three times as many as might be expected based on familiarity alone. While this success has yet to be reflected in strong revenue growth (like many luxury brands, Burberry suffered from the weak economy in 2012), its strong digital presence positions it well to take advantage of the growth of digital commerce.

PLANNING FOR SOCIAL MEDIA SUCCESS

Major marketers the world over are beginning to recognize that social media is no free lunch. Social media requires the same discipline and rigor as traditional media—and perhaps more—if it is to pay its way. In addition, social media cannot be considered in isolation from other media; it must be integrated with them.

With this in mind, Coca-Cola has adopted the idea of "liquid and linked" cross-media marketing: "liquid" because the content needs to flow seamlessly from one media channel to the next, and "linked" because the impression conveyed about the brand must remain focused on what the brand stands for. Using the same core idea in different media, but particularly digital, allows the brand's storytelling to become dynamic, evolving over time to create deeper audience engagement and participation.

FOCUS CONTENT ON THE MEANINGFUL DIFFERENCE

In spite of a reticence to feature brands too strongly in viral videos, one of the most successful viral campaigns is actually nothing more than an old-fashioned product demonstration. BlendTec's videos featuring the company's CEO Tom Dickson were an early viral success. Starting in 2006, a series of videos were posted to YouTube showing Dickson blending a wide variety of items, including 50 marbles and a garden rake. Since then, the formulaic video series has shown the BlendTec get the best of iPhones, golf clubs, and more. By the end of 2012, the brand's YouTube channel had received 220 million views, had over 500,000 subscribers, and is credited with significant business success.

In spite of its single-minded product focus, the BlendTec videos are engaging and do not come across as a hard sell. There is an innate fascination in watching something valuable like an iPhone get chopped to dust in a matter of seconds. And the videos also make the power and durability of the blender an inescapable part of the experience. When the product is what makes your brand meaningfully different, then you need to find ways to highlight what it does well. Dyson took a leaf out of the BlendTec playbook when it created a video showing a balloon being sucked through a series of Air Multiplier fans. I doubt many people really want to know the physics behind the fan's unique design, but the video engagingly demonstrates "inducement and entrainment" in action. The video has attracted over 2.5 million views and spawned an iPhone app.

CREATE SHARABLE EVENTS AND CONTENT

What about a brand whose meaningful differentiation is more intangible? Coca-Cola's Happiness Machine provides a classic example of a video that captures the essence of what the brand stands for. In addition, the video has proven liquid enough to morph into a TV commercial and inspire other events.

Filmed with hidden cameras, the original video shows students at St. John's University in New York astonished by a Coke vending machine that dispenses two bottles of Coca-Cola instead of one, flowers, pizza, and even a giant sub sandwich. The expressions of surprise and enjoyment on the students' faces are a delight to watch, and they are entirely in line with Coca-Cola's promise to "Open Happiness." Created by Definition 6, the video was originally made as a piece of digital content that the company hoped would go viral. It was well received, and based on research conducted by Millward Brown, Coca-Cola subsequently adapted the video for TV. The Happiness Machine idea was then replicated and adapted in other countries with equal success.

"Personality becomes extra important given the conversational and relatively unscripted nature of social media," says Tom Buday, Nestlé's head of marketing and consumer communication, and he points to the company's Skinny Cow brand as a case in point.[13] Skinny Cow offers low-fat ice cream and candy and believes that "looking good and staying healthy doesn't mean a life without indulgence." The brand has grown steadily, and Skinny Cow has now attracted over 600,000 likes on Facebook. Buday notes that an important aspect of maintaining the brand's online persona is the list of do's and don'ts that the company has created to ensure a consistent tone of voice over time.

Fitting well with Skinny Cow's personality and acting as a catalyst for comment are the brand's Perfect Cup Events. Recognizing that 80 percent of women wear the wrong bra size, these events offer free fittings, vouchers for a new bra, and, of course, free ice cream cups. Judging by the commentary on blogs, Yelp, and elsewhere, these events are both well organized and well received. The Perfect Cup Events are a great example of using events in combination with social media to create buzz around a brand. The juxtaposition of bras and ice cream has inherent talkability, and when the events provide a positive experience, people are bound to talk them up, usually to folks of like mind.

■■■■ SEED YOUR CONTENT WIDELY

Very few "viral" videos achieve significant viewing success, and almost none of them go truly viral. To be truly viral, each person who received the video would have to pass it on to more than one other person, and that almost never happens. If you want to hit critical mass with a video (generally considered to be 1 million views or more), then you need both great creative and wide dissemination. When the pop video "Gangnam Style" by Korean rapper Psy took off in 2012, it rapidly achieved worldwide fame (with over 1,250,145,678 views by the end of January 2013), but it took more than a catchy tune and a fun video to achieve that success. Psy's music company YG Entertainment is reported to have invested in organically growing an engaged audience, with 2.5 million people following its various YouTube channels, so that when the right song came along, they had a large platform on which to seed the campaign and ensure exposure.[14]

■■■■ MEANINGFUL DIFFERENCE AND VITALITY

Successful marketing is not simply a matter of creating buzz. While it is important that a brand be seen as part of the contemporary culture, it is just as important to ensure that what makes the brand meaningfully different is kept salient in the minds of its target audience. Gordon Pincott puts it this way: "A major role for marketing is to keep the brand and the idea of that brand readily accessible at times when people are looking to fulfill a need or make a purchase."[15]

If your marketing fails to focus attention on what is meaningfully different about the brand, then it is wasting an opportunity. The real challenge of marketing remains to create engaging content that highlights the brand and what it stands for and then disseminate that content as widely as possible. The advent of social media has not changed this remit. If anything, it has simply made it more imperative that marketers do their job well.

Questions to Consider

1. Is your brand first to mind when a particular need arises? What is it that comes to mind when people think of your brand? What sense of the brand are you trying to foster, and is that how consumers actually experience it? Does the brand inhabit the same spaces where its target audiences spend their time? Does it add something to that experience? Is the brand perceived as popular and contemporary?
2. What opportunities exist for your brand to play a role in and be welcomed into people's lives? Make sure the brand is present when and where people might want to encounter it. While specific communication strategies may differ, ensure the brand conveys the same essence across all touch points.

Chapter 13

Affordability at a Price Point

Apart from the quality of the product or service, affordability is the most important variable in the marketing mix. Finding ways to make it easier for people to buy your brand will help entice new buyers, but if the balance between price and value perceptions is off, then the brand will end up losing margin and enter a death spiral. As a brand's income declines, funds for research and development and marketing are choked off, reliance on price promotion to drive volume grows, and the brand becomes uncompetitive. To avoid this demise, you need to make the brand more affordable—not simply cheaper. A savvy marketer will be able to manipulate people's perceptions of value in order to maximize the income stream from a brand at a specific price point or find ways to make the brand more affordable without lowering margins.

PRICE DECREASES AND VOLUME SALES

A few years ago, I conducted an analysis based on a set of 354 consumer packaged goods brands for which we had volume and value market share data across the course of two sequential years. *Volume market share* is the number of units sold by each brand divided by the overall category sales. *Value market share* is the dollar sales of each brand divided by the overall category dollar sales. The brands in this analysis came from a wide variety of categories and countries, with data being sourced from our clients or directly from sales measurement companies.

From the two types of market share data we were able to derive another important measure: the *relative price index*. If you divide a brand's value market share by its volume market share, you get a measure of price relative to the category average. In other words, this number tells you whether the brand was sold at a higher or lower price than the average. Table 13.1 shows a summary of what I found.

Table 13.1 On average volume share holds steady but relative price falls

	% Change in Volume Share	% Change in Value Share	% Change in Relative Price Index
Median across all cases	0.0%	−1.1%	−1.3%
Increased	43%	37%	34%
Flat	7%	10%	5%
Decreased	50%	53%	61%

So what do these data tell us? The median volume share change was zero, indicating no change on average from year to year. The number of cases in which volume share increased was similar to the number of cases in which it decreased. This confirms what we have seen before—overall market share changes very little from year to year.

Value market share shows a slightly less positive picture. The median value market share fell 1.1 percent year on year, and the balance shifted slightly further toward decreases rather than increases. How can this be if volume share change is constant? Because the median relative price fell 1.3 percent. The proportion of cases in which the relative price index decreased is almost double the number of cases in which it increased. Some brands improved their price premium relative to the competition, but most held their relative position or lost ground.

How does this happen? Most brands prioritize maintaining volume market share over maintaining their relative price point. Often the rationale given is that reducing prices and maintaining volume staves off expensive layoffs and factory closures. Unfortunately, profits tend to suffer under this approach, and the company has to pay the price in the longer term, as the U.S. car industry discovered. If you want to maintain a healthy business, you have to find ways to grow profits. Later I will look at some of the implications of this analysis more closely, but first let's consider several reasons why companies focus on volume to the detriment of price.

1. Consumer Desire for Cheaper Prices

I would argue that the data reviewed in the previous section is representative of most established product categories. There is long-term downward pressure on prices simply because of consumer demand—everyone likes a good deal. Ardent brand fans will eagerly buy their brand if it is cheaper than normal, and woe unto the marketer who by accident or design encourages buyers to expect deals and discounts, since the expectation of future deals will simply encourage them to shop around or wait for a deal to become available. Some shoppers will do this anyway; in all categories, a proportion of buyers typically place more importance on getting the best price than on which brand they buy. The proportion varies according to the investment required to buy within the category and the perceived risk that a brand might turn out to be a "bad" choice. The more likely people are to find a brand in the category to be meaningfully different, the less likely they are to value price over brand choice.

2. Targeted, Not Generic, Price Promotion

Marketers need to be very clear about where, when, and how to use price promotion. All the evidence suggests that price promotion should be used tactically and with very specific objectives in mind. Identifying the target audience is particularly important. Generic price promotion will drive volume, but it may well come from price-sensitive shoppers who only buy the brand when it is on sale. When the marketer attempts to end the discount, these price-sensitive shoppers will choose another, cheaper brand instead. Meanwhile, the marketer leaves money on the table by selling discounted product to people who would have bought it anyway.

All academic and industry research suggests that price deals and discounts are a brand's equivalent of crack cocaine. Once you start using them frequently, it is really difficult to stop. Worse, pricing is such an obvious variable in the marketing mix—for competitors, retailers, and consumers—that changes are unlikely to go unnoticed and ignored in the long term. Unless they are confident in the strength of their brand, competitors will feel forced to match promotional price decreases, leading to a price war in which no one wins. Lowering a brand's price is the obvious way to drive more volume, but businesses should do everything they can to resist that downward pressure. Even if you are a low-cost producer, there is a price point at which your margins are so low, you might as well give up and put your money in the bank.

3. Focusing on the Wrong Metric

The fundamental problem is that most companies place their day-to-day emphasis on driving volume, not profit. When senior management defines the measure of success as driving top-line growth, they automatically focus attention on how much is being sold—not on how much value is being created. In my experience, management objectives and incentives tend to reinforce this focus. The real issue is that sales are easy to measure; the data often exist in real time. It is a lot harder to measure the profit that results from those sales, and such data are often only available after the event (although this availability is changing as new, smarter sales and pricing algorithms are deployed). Every idiot knows that the way to sell more is to drop the price. The data prove it. In the market share change dataset described previously, decreases in the relative price of a brand correlate with increases in volume market share. But the correlation between changes in relative price and value market share (a measure which accounts for both how much is sold and at what price) is weaker. Selling more at a lower price does not always improve revenues, let alone profits.

Importantly, premium brands are not necessarily at a disadvantage to cheaper ones simply because they cost more. Provided the brand is able to justify its premium, it can outperform the cheaper competition. Analysis of the consumer packaged goods brands dataset reveals that premium brands are marginally more likely to grow market share from year to year than value brands. On average, a brand which charged 50 percent more than the category average price grew volume by 1.9 percent year on year. The average growth for a brand charging 25 percent less than the category average was –2.2 percent. This finding should come as no surprise. Companies invest in innovation and marketing with the express purpose of achieving higher growth rates while still justifying a premium price.

▬▬ JUSTIFYING A PRICE POINT

According to a meta-analysis reported in *Empirical Generalizations about Marketing Impact* published by the Marketing Science Institute, "Long-term sales growth for a brand is derived mainly from category growth."[1] Such growth is hard to come by in many mature categories. However, the value of a brand isn't determined simply by sales; it is also determined by profitability.

If you can command a premium price, then you can increase the value of your brand without significant volume growth. This is a considerable advantage. The consulting firm McKinsey finds that a 1 percent increase in average price (assuming volume remains stable) improves operating profits by 8 percent.[2] That is three times the increase generated by a 1 percent increase in volume.

A 1 percent decrease in price has the opposite effect. McKinsey concludes that a strategy that cuts prices to drive volume is "generally doomed to failure in almost every market and industry."[3]

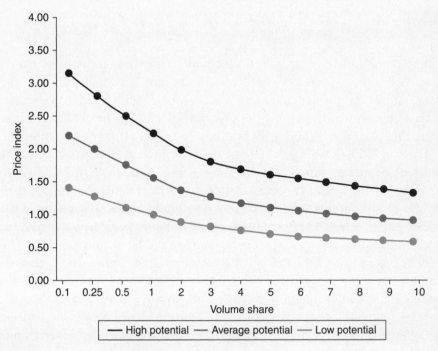

Figure 13.1 High potential brands command a price premium

Source: Millward Brown analysis of BrandZ and IRI sales data

So what can justify a price premium? *Perceived differentiation*. Provided brands are seen as desirable choices, they can thrive at higher price points when they are well differentiated—that is, when they have a meaningful, relevant, and valuable point of difference. Figure 13.1 shows the results of a Millward Brown analysis of brand equity metrics against IRI volume share and price data. Better-differentiated brands—those that do better at converting people through meaningful differentiation—are able to support the same volume at higher prices.

The fundamental challenge of affordability is to grow sales at a set price point and avoid using price itself to drive volume.

Figure 13.1 also illustrates the trade-off between volume share and premium prices. The choice between whether to seek growth or whether to command a price premium depends on the brand's financial model and how easily it can extend its appeal to a mass market. Some exceptional brands can actually manage to do both. Starbucks, Tide, and Pampers have all managed to grow market share while simultaneously maintaining the perception that they are different and commanding a price well above the category average.

A STRONG BRAND STORY CAN JUSTIFY A PRICE PREMIUM

As noted previously, companies invest in innovation and marketing in order to sustain a price premium and grow the brand. The challenge is to find something about the specific brand that resonates with enough consumers so that they are willing to pay the premium. That was the challenge facing Bega, the Australian brand leader in the everyday natural cheese category.

Early in 2010, the everyday cheese category in Australia was highly competitive and facing some tough challenges. The domestic brands in particular were facing a new, low-cost competitor, Kraft, which had entered the natural cheese market in August 2009. The brand leader, Bega, was in particular jeopardy. Kraft's packaging looked similar to Bega's, but the product was considerably cheaper, as were the store brands that commanded a substantial proportion of the market. Faced with rising commodity prices and deep discounting, Bega had to come up with a way to maintain its value market share and its position as Australia's favorite cheese brand.

Bega's agency, Clemenger BBDO Melbourne, identified a growing trend of consumer skepticism toward big business; people were increasingly being drawn to locally produced goods. A successful 2008 campaign had started to tap into this desire by highlighting the origins of the cheese: the country town of Bega in New South Wales. As an Australian Effie Awards case study notes, the new campaign built on that foundation: "In this campaign, we leveraged our biggest advantage over the U.S. giant, Kraft. We had provenance and real people on our side. Bega was Australian, local and proud of it, and our farmers could express this better than any script we could ever write."[4]

The "Real Farmers" campaign made heroes of the farmers who produced the cheese and their families. The video ads were unscripted interviews shot with a hand-held camera to give a more authentic and down-to-earth feel. They were shown on TV and in movie theaters, and print and outdoor advertising supplemented the campaign.

The campaign resonated with Australian consumers. Millward Brown tracking found message recall to be high and enjoyment of the ads to be in the top 25 percent of all Australian advertising. In a year of static category sales, Bega grew its revenues by 3.7 percent. Bega's value market share increased from 11.7 percent in 2009 to 12.1 percent in 2010 in spite of continued aggressive competition. The agency calculated that the campaign achieved a return on investment of $3.67 for each $1 invested.

The ways in which a brand can justify its price premium are as varied as the brands themselves. Some rely on tried and tested value for money messaging. For instance, in an apparent response to the launch of the Dollar Shave Club in the United States, Gillette dramatized how long a single Fusion ProGlide cartridge would last. The razor's ad shows a man traveling the world for five weeks and concludes, "One world, five weeks. The only thing that did not change was my razor." Other brands simply ignore price as a variable and focus on selling the benefits of the brand. Apple does not directly justify its pricing through advertising; it simply demonstrates what it is like to use the product. What Apple does instead is to focus on aggressive cost control within its business model to ensure that a like-for-like product from a competitor (should such a thing really exist) will be unlikely to command a significant price advantage.

■■■■ MANAGING BRAND PORTFOLIOS TO MAXIMIZE APPEAL

Many companies build brand portfolios consisting of products at different price points. For example, most carmakers offer a variety of models—BMW has the 1, 3, 5, and 7 Series, while Honda has the Accord, the Civic, and the Jazz or Fit. Johnnie Walker has Red, Black, Double Black, Green, Blue, and Gold Label options. In the United States, Purina, a major manufacturer of pet food, offers their base brand, Purina Dog Chow, as well as more upscale and specialized options such as Purina One (available in grocery stores) and Pro-Plan (sold through pet specialty channels).

Such an approach is often successful but is complicated to manage. It requires that the company understand and delineate the strategy for each individual brand in relation to the needs and values of various consumer segments. The company must identify the meaningfully differentiated experience that justifies the price of one brand over another, and it is then incumbent upon the brand to ensure that the value proposition of each product is clearly communicated to consumers.

The relationship among the various brands within a portfolio must be thought through and managed. If consumers are expected to move up or down the brand ladder as their life stages change, then the connection between the brands must be clear in order to keep consumers "within the family." In one of our ValueDrivers workshops, we helped the local team of a well-known brand identify what made its more upscale offering meaningfully different from its popular brand. The popular brand dominated its category, and most consumers found it to be meaningful and different enough. They saw the more expensive product as being the same brand and struggled to identify what made it worth paying 30 percent extra. In the end, we identified a platform that helped justify the brand's premium and a way for it to appeal to target consumers with a particular mindset.

▄▄▄▄ BRANDING DOWN: EXTENDING APPEAL THROUGH LOWER PRICE POINTS

Companies around the world now recognize the need to "reach down" if they are to better serve the 2 billion consumers entering the middle class in developing economies, but this principle is not limited to those economies. In developed economies hard-hit by the Great Recession, the "have-nots" represent a large market that would welcome more affordable products. Particularly when trying to extend a brand's appeal to lower-income consumers, marketers have to find ways to make the brand accessible without undermining its perceived value or actual margin. This is a delicate balancing act. In some cases, it requires selling smaller quantities of the product. In others, the manufacturer might provide financing options that make it easier to buy. For example, in Chapter 5, we learned how Coway boosted sales during a recession by renting its products as well as selling them.

In developing markets, it often happens that even when multinational companies reduce production and distribution costs as much as they can, the absolute cost of their brands remains high in relation to the income of the local population. In these instances, the brands can still be made available to aspiring consumers using creative payment options. For example, in India, Levi's offered consumers the option of buying jeans on a monthly installment plan, thereby making the brand seem more affordable without actually reducing the absolute price. This approach attracted a large number of new consumers who had strong positive views about the brand. Previously, these people would have settled for a $10 pair of jeans bought in the local market, but now they could afford the brand they really wanted with three monthly, no-interest payments. The perception that the brand was too expensive dropped, and the proportion of people claiming to have most recently bought a pair of Levi's more than doubled from one year to the next.

▄▄▄▄ BRANDING UP: GENERATING MARGIN THROUGH RARITY

Not all brands will benefit from making themselves more affordable. Luxury brands—like Dior, Louis Vuitton, and Hermés—have always benefited from the cachet that comes from being available to the wealthy few, and some premium brands have been successful in leveraging scarcity and rarity to their advantage. Limited editions are a tried and tested means of creating renewed interest in a brand and generating a good financial return at the same time.

My current car is a limited edition. In 2008, only five thousand VW R32's were imported into the United States, and that was it—no more were produced. The addition of 4Motion all-wheel drive, direct-shift gearbox (DSG) transmission, and a 3.2-liter V6 engine helped the R32 stand out from the regular Golf crowd and raised it to a rank alongside higher-end sport sedans. While lacking the same eye-candy appeal, the R32 became the flagship in the Golf lineup in the same way the R8 was for Audi. More important from the manufacturer's viewpoint was the fact that the R32 set a new price benchmark against which people could judge the GTI, previously the most expensive model in the Golf lineup. With a base price of $32,990, the R32 provided an "anchor" price, one that encouraged people to choose the 2-liter turbocharged GTI for $9,000 less and then load it up with high-margin extras. In 2012, VW again deployed the same strategy with the introduction of five thousand Golf R's, leaving today's buyers to ponder the question: Is the R special enough to pay $3,995 more than you would for a 200-horsepower, front-drive Golf GTI Autobahn? Of course, if you don't buy the R, then it will probably be another four years before you can reconsider that decision.

Creating new value-added options to create higher margins and provide a benchmark price for cheaper options is not a strategy limited to the car category. As noted previously, Johnnie Walker has an extensive lineup of whisky blends at different price points and has generated 20 percent growth of its super-premium brands by targeting celebrations like Diwali and Chinese New Year with the sale of special gift packs.[5] Special blends also create the potential for premiumization. Johnnie Walker XR21 is aged for an extended 21 years and comes in unique packaging that helps differentiate it from the existing blends.

LOW PRICE AS IDENTITY

This chapter would not be complete without a consideration of value brands. Ryanair, the leading low-cost airline in Europe, has built its strength on its ruthless pursuit of low cost. The budget carrier not only flies more passengers than British Airways, but it is also more profitable. As part of its quest to drive down prices, Ryanair created new routes to smaller airports to save money on landing fees. This has given the brand another point of differentiation—it serves regions traditionally neglected by other airlines—and provides the company with significant leverage over the owners of those airports and local tourist industries.

To sustain low prices over the long term requires a commitment that has consequences for every aspect of the way a company works. For Ryanair, this move involved rethinking not only the routes, but also the role of the cabin crew, the strategy for managing onboard and in-hold luggage, the algorithms used to price tickets, and so on. Ryanair also recognizes that its single-minded pursuit of low costs will inevitably alienate a portion of the potential customer base. For example, some people will not accept the lack of assigned seating as a fair trade-off for a lower fare. And while many people may appreciate the transparency of Ryanair's à la carte approach to pricing (for priority boarding, leg room, food, and luggage), others will be put off by this practice.

■■■■■ LOW PRICES DO NOT PRECLUDE QUALITY OR A POSITIVE EXPERIENCE

SuitSupply is a men's fashion brand. The brand is targeted at younger men and seeks to make the suit an affordable and popular fashion item. While the garments—which also include shirts, shoes, and other accessories—are made with traditional craftsmanship and quality, the company's business model is designed to do so at a low cost. According to an article in the *Wall Street Journal*, a $614 suit from SuitSupply matched a $3,600 Armani in quality and outperformed designs from J. Crew, Hart Schaffner Marx, Target, and H&M.[6]

In order to ensure low prices, SuitsSupply is vertically integrated. Unlike other better-known luxury brands that are often distributed to retailers through middlemen, SuitSupply ships product directly from its own factories to the brand's stores. The stores themselves are usually found on the second floor of a building, not at street level. No doubt this placement saves hugely on rent, but it also saves staff time by discouraging people who are simply browsing with no intention to buy. With suits starting at less than $500, a tiered price structure encourages people to trade up, although they rarely break the $1,000 mark. The design of the stores separates them from the traditional, fusty feel of many retailers. The Soho, New York, store that I visited has an open and airy feel, with white walls and plenty of natural light. A tailor's workstation is immediately visible on entering the store, and attentive staff receive bonuses not on the amount they sell, but on follow-up customer satisfaction gathered via email.

The whole SuitSupply experience has been engineered to make the process of buying a suit positive while at the same time making it much more affordable. But lower prices can result in lower margins, and this therefore creates a need to be a volume player in the market. Walmart makes high levels of absolute profit by balancing lower margin with enormous scale; Aldi keeps the price of its product low by selling only one item in each product category. Given the rate at which SuitSupply is opening stores, it makes me suspect that they are well aware of the need to generate volume in order to boost profits.

A BALANCING ACT

As Ryanair and SuitSupply demonstrate, there is always a market for low-priced brands, but to make money these brands must drive volume and fine-tune their business models to remain low-cost providers. For most brands, manipulating price is a double-edged sword. Cutting prices will drive more volume but at the same time will encourage people to look for more discounts in the future. Unless you are a low-cost provider, this is a dangerous route that is likely to undermine profit growth; it is far safer to sell the brand at a profitable price point and justify that price through meaningful differentiation. However, you should still seek mechanisms to make your brand more accessible. Monthly payment plans, zero interest, and loyalty and layaway schemes all offer ways to extend access to a brand. And if the price point still remains out of reach to most consumers, then marketers should seek to brand down, keeping in mind the need to clearly differentiate the pricing tiers within the brand offering.

Questions to Consider

1. Is there an agreed-upon pricing strategy for your brand? Do you know that the current price point is optimal and generates the highest margin? Are you doing everything you can to make it easy for people to buy your brand at that price point (without resorting to discounts)? Are you structuring and bundling the offer, shaping the architecture, defining the sizing, and looking at payment terms in innovative ways to maximize your market at the given price?

Chapter 14

Extendability

For a strong brand, extendability is how you reap the real benefits of your investment. A meaningfully different brand can extend to new usage occasions, product categories, and geographies without every step of the process needing to be mapped out from first principles. Adaptation will no doubt be necessary, but knowing that what the brand stands for already resonates in one category will make that adaptation easier. Brand extension should not be limited to product lines and geographies. Licensing is an attractive option for companies that aim to boost revenues with little immediate risk. Some brands—like Virgin, Disney, and Ikea—possess the strength to compete effectively in widely differing categories with little functional similarity. For others, in situations in which strong brands have become lost inside a bigger conglomerate, spinoffs may offer the best way to create increased value.

◾◾◾ REACH UP, DOWN, AND WIDE TO GROW

Unilever aims to grow its sales to $106 billion in 2020 (double its sales in 2009) and expects the majority of that growth to come from extendability. While Unilever's strategy encompasses all the brands in its portfolio, the same principles apply to individual brands. 70 percent of Unilever's revenue growth is predicted to come from continued market development (with the remainder split between market share gain and white space—that is, completely new products).[1] The company's strategy for market development can be described as: "Reach up, reach down, reach wide." Unilever plans to reach up by developing premium brands that will appeal to the "have-lots." The intent is to garner the same growth and margin improvement across the whole portfolio that is already apparent in the hair care and ice cream categories. They plan to reach down and develop value brands, entry-level products, and pack sizes that address the needs of the "have-nots." And they plan to reach wide by expanding into new geographies, segments, and retail channels.

Bringing it back to the brand level, one only has to look at the expansion of the Dove brand to recognize how powerful extendability can be as a driver of growth. In 1990, prior to its global expansion, the Dove brand generated sales of just under $400 million from its bar and dish soap product lines. Over the next four years, the brand was launched in 55 new countries and has continued to expand its footprint since. In 1995, Dove body wash was launched, followed by deodorant in 1997, body lotion in 1998, facial cleansers in 1999, hair care in 2002, an anti-aging range in 2006, and Dove for Men in 2010. In 2011, Dove became Unilever's first $4 billion brand.[2] The next challenge for Dove will be to fill out its own portfolio around the world and to address markets like China, where the brand has struggled to establish the same meaning that it has in the West.

As I explained in *The Global Brand*, extending a brand to new countries and cultures is not without risk.[3] Like Dove, many brands struggle to establish meaningful differentiation in the context of new cultures. However, launching an established brand in a new country is considerably less risky than trying to launch a completely new one. New mistakes may be made, but at least you have a blueprint for what worked the first time around to guide your actions.

HOW STRETCHABLE IS YOUR BRAND'S MEANINGFUL DIFFERENTIATION?

How strongly a brand can benefit from extendability depends in large part on its meaningful differentiation. Gordon Pincott notes: "Brands based upon an ideal or with a purpose that provides scope are in a stronger position to stretch across different categories. The frame of mind of the brand owner and the consumer makes it more amenable to see this brand in different places. At the same time, a clearly defined purpose will highlight specific categories where the brand could play a role and place other categories off limits."[4]

If what the brand stands for is limited to a functional benefit or a product category, then its extendability may be limited. In the past, McDonald's has struggled to extend beyond fast food. Coca-Cola has been content to demonstrate its mastery of product extension within the soft drinks category. Of course, one element of successful brand extension is a proven track record. Dove could have remained a humble soap bar if it had not managed to exploit its original moisturizing benefits to enter new categories.

FUNCTIONAL BENEFITS

From a consumer viewpoint, functional credentials play an important part in determining what is regarded as a "good" extension. Dell and soft drinks don't sit easily together for most people, nor do Levi's and skin care or Sony and motor fuel. However, a functional benefit does not necessarily limit a brand to one category.

Yunnan Baiyao is a well-known brand in China, where it leverages its traditional Chinese medicine heritage to compete effectively against multinational brands like Colgate, Crest, and Johnson & Johnson's Band-Aid. Yunnan Baiyao literally translates as "the white medicine from Yunnan." Originally created in 1902, the herb-based powder gained a strong reputation for its curative properties, particularly its ability to stop bleeding. Over a century later, in 2005, Yunnan Baiyao introduced its sticking plaster product with the same medicinal properties. Given the brand's heritage, the application made perfect sense to Chinese consumers. Three years later, the brand transferred its medicinal heritage to oral care with the creation of Yunnan Baiyao toothpaste. The promise to clean teeth, keep gums healthy, and stop bleeding has proved attractive. In 2012, BrandZ found Yunnan Baiyao to be a meaningfully different brand able to justify a price premium compared to the bigger and more familiar multinational brands.

TRANSCENDING FUNCTIONAL BENEFITS

What does Ikea stand for? Is it just cheap furniture? Or is it something more fundamental?

Ikea's founder, Ingvar Kamprad, is a notoriously frugal billionaire who supposedly said about himself, "I am stingy, and I'm proud of the reputation." (Kamprad reportedly drives a fifteen-year-old Volvo 240 GL.)

Frugality is central to Ikea's success. Furniture sold in the store is designed to meet target price points and serve its function well. If the Ikea brand were defined by furniture, then its growth curve would be confined by how many people are willing to buy flat-pack furniture from big-box stores. But position the brand as "smartly frugal," and the addition of flat-pack houses, budget hotels, and student accommodations to the company lineup makes perfect sense. Strangely, it has been reported that the hotels will not carry the Ikea name (unlike the hotel in the company town of Almhult).[5] I can't help wondering why. People might be surprised to see an Ikea hotel, but they would know exactly what to expect from it. It will not be a luxurious experience, but it will be stylish and economical.

What about a town designed and built by an Ikea company? LandProp, part of the Inter IKEA Group, is busy developing a 26-acre, near-derelict site just south of the Olympic Park in London into a waterside city neighborhood.[6] The project, known as "Strand East," will consist of 1,200 new homes, commercial space (including a "creative zone" for industry), local shops, cafes, restaurants, and a 350-bedroom hotel. Plans call for about 25 percent of the development to be open space, including a riverside park that will run along the western edge of the site and overlook the River Lee Navigation. As of the time of writing, the future town is marked by a slender, 40-meter tower constructed of wooden lathes. At night, six hundred LED lights create a beacon, signaling big things to come. Strand East is an ambitious project, but with planning permission in hand, its main construction activities will likely start by 2014.

◼◼◼ PROTECTING YOUR CURRENT BRAND ASSET

Brand extension is not an endeavor to be undertaken lightly. Unless managed carefully, the potential exists to waste time and money, alienate users, and dilute the existing brand's equity. Four basic principles are worth bearing in mind when considering any brand extension.

1. Make Sure Your Business Model Creates Competitive Advantage

In its home country of the United States, Walmart is a strong brand, indexing above average as meaningful, different, and salient. But as Walmart's management has learned, strength at home does not always translate to strength abroad. Hoping to gain a significant share in Europe's biggest market, Walmart launched in Germany in 1998. American management practices soon ran headlong into conflict with local labor law and workplace culture. These problems were compounded by a lack of understanding of local shopping needs and strong competition from local discount brands like Aldi and Lidl. In 2006, Walmart announced that it was calling it quits and sold its stores to the Metro Retail chain. The withdrawal was expected to cost the company $1 billion.[7] Unfortunately, the Walmart Germany debacle is not unusual. Retail is one of the businesses most sensitive to differences in local culture, but retailers often fail to appreciate that fact.

While on a much smaller scale than Walmart's entry into Germany, back in the United States, Procter & Gamble has extended its Mr. Clean brand from household cleaner to car washes. Conceptually, if "a clean shine delights" around the house, then it should delight on your car as well—or so Procter & Gamble hopes. The first Mr. Clean car washes opened on a pilot test basis in 2007 in Cincinnati. In 2009, the company acquired the franchise assets of Atlanta-based Carnett's Car Wash, with the previous owners now heading up the Mr. Clean Car Wash franchises.[8] In theory, the fragmented car wash industry does represent a good opportunity for a big player, and it is obvious that Procter & Gamble have done their best to find ways of improving the car wash experience (with a satisfaction guarantee, comfort lounge, and recycled water); however, it is not clear that Mr. Clean's iconic muscle-bound genie is going to be able to work his magic in this business. As of 2012, the majority of the Mr. Clean car washes remain located in the greater Atlanta and Cincinnati regions.

I suspect that the fundamental issue facing Procter & Gamble has less to do with brand strength and more to do with its business model. Unless the Mr. Clean car washes fundamentally change the economics of the business, then they are starting on a level playing field in competition with the thousands of local car washes already in existence. Brand strength will only carry things so far, and so although it is well known for its strong cleaning heritage, Mr. Clean is far from the household name it once was.

2. Protect Your Brand's Meaning

Starbucks is a classic example of both the risks and the rewards of extendability. The risks have come from a lessened focus on the core brand offer. The rewards have come from extending that core offer to new geographies.

Let's look first at the expansion of physical availability. Starbucks was founded in Seattle in 1971, and it was not until 1987 that the brand entered Canada. The company went public in 1992, and the late 1990s saw expansion into Japan, the United Kingdom, and China. Expansion continued in the new century with stores opening in Brazil and Russia in 2006 and 2007, respectively. As of 2012, Starbucks had more than 17,400 stores worldwide, including three in Mumbai, India. At its biennial investors' conference, CEO Howard Schultz announced the goal of having more than 20,000 retail stores on six continents by 2014. Currently Asia represents the company's fastest-growing market for retail stores, with China expected to become Starbucks' largest market outside of the United States in 2014.

But the current expansion marks a recovery from an earlier setback. In 2007, the prospects for further significant growth in the United States had seemed somewhat lackluster. Starbucks was thought to have lost its focus on the coffee experience. Machines automatically churned out espresso with the push of a button instead of the barista making each shot by hand. The smell of breakfast sandwiches competed with that of coffee. And the cozy feel of a neighborhood cafe had given way to a more bland experience. These problems were documented by none other than Howard Schultz in a memo titled, "The Commoditization of the Starbucks Experience" in February of 2007.[9] In spite of successful expansion abroad, the memo seemed to confirm investors' worst fears. From a high of $39 the previous year, Starbucks' share price had declined 40 percent, sliding still further in 2008 to bottom out at around $10. It was not until 2011 that the share price returned to and then surpassed its previous high, after revenues and same-store sales had returned to growth in 2010.

While the Great Recession undoubtedly revealed the brand's weakness, Schultz ascribes the real cause of that weakness to unfettered growth: "The big issue I think was that growth is not a strategy, it is a tactic, and if growth becomes a strategy I don't think it is an enduring one. I think growth covers up mistakes."[10]

As told in *Onward*, the story of the company's transformation, a first step to revitalize Starbucks was taken when Schultz put the brakes on expansion. In the United States, 600 stores, representing 5 percent of the company's total stores, were closed. Notably, 70 percent of them had only been open for three years, but in the quest for growth, they had been located too close to other Starbucks stores, reducing traffic for each one. Cost reductions, process improvements, value offerings, loyalty schemes, and store makeovers all helped return Starbucks to a successful growth track in the United States, and for the first time, Starbucks started advertising, reminding loyal customers of what the brand stood for.

In preparation for future extension plans in 2011, Starbucks also took control of the distribution, marketing, and promotion of its packaged goods brands back from Kraft and removed the word "coffee" from the brand's logo. In a January 2011 post on the Starbucks blog, CEO Howard Schultz openly admitted that the new design, while harking back to the brand's heritage, is intended to "give us the freedom and flexibility to explore innovations and new channels of distribution."[11]

While the logo change created significant debate, one other change seems to have gone unnoticed. In 2011, 63 percent of international stores were licensed operations, a direct contrast to stores in the United States, of which 62 percent were company-operated.[12] This represented a big shift in strategy for a brand that had traditionally only ceded control in locations such as airports and hotels, where another business controlled the retail space. While licensing is a great way to expand a business without incurring capital costs, it also requires stringent training and quality control if the brand experience is not to be undermined. It will be interesting to see how well

Starbucks manages to keep its meaningful difference alive without direct control, but Schultz is the first to recognize that holding true to the company's ethos is critical. In an interview for the UK's *Guardian* newspaper, when asked whether anyone else could run the company, Schultz replied: "The next CEO of Starbucks should come from within, and it is my intent to ensure that happens because it is very difficult to teach the imprint, the values and the culture and the complexity of what we do to an outsider."[13]

3. Make Sure You Can Deliver on Your Promises

Books, planes, and banks: they can all be found in Virgin's eclectic company lineup. But did you know that Virgin also offers vacations, wine, and health care? Superficially, there seems little rhyme or reason to the industries served by the Virgin brand, but that is far from the truth; in fact, the brand's values and entry strategy are consistent.

CEO Richard Branson is a consummate entrepreneur, and his strategy for brand extension is well documented—identify categories in which the consumer has traditionally received a poor deal and the incumbents are big and complacent. Branson has been quoted as saying, "We look for opportunities where we can offer something better, fresher and more valuable, and we seize them."[14] This approach makes Virgin the challenger brand in every category it enters, and the brand's commitment to great service, irreverent attitude, and underdog status have served Virgin well in most cases. The widespread application of this strategy is enabled by the company's business model. Branson calls Virgin a "branded venture capital organization."[15] He manages the business under the Virgin name, and investors underwrite the start-up costs. This approach allows Virgin to take on a wide variety of different challenges, some of which have proved less successful than others.

Virgin's entry into the computer, cola, vodka, and personal care categories are numbered among the brand's failures over the last two decades. Virgin Cola's long, drawn-out demise came about because Virgin underestimated the power of the incumbent brands and their ability to fight back. The computer debacle happened when Virgin ended up selling an unexceptional product designed by another company. For many brands, a well-publicized failure would be a black eye at best and a cause for resignation at worst, but Branson sees failures like these as learning experiences. In a *Fast Company* article published in 2004, Branson states, "We had decided that we'd never get involved in a company unless we could control it and protect the Virgin brand name."[16] So long as Virgin is living up to its challenger status, Branson seems unconcerned with failures. He seems to believe it is okay for a company to go down swinging, particularly if it garners some good publicity in the process.

4. Know When to Stop

I have to admit that when I first heard about the launch of Dove Men+Care, I wondered, albeit fleetingly, if it was a joke. How could the brand that famously espoused the Campaign for Real Beauty—single-mindedly targeting women—think that launching a men's line was a good idea? I am sure the business case made sense, and the brand will generate more revenue in the short term. However, I worry about the long term and the ambivalence that the line will create among those who bought into the Dove philosophy when it stood for revealing a woman's inner beauty. Only time will tell if my concerns are correct, but I believe that all too many successful brands get carried away with extension and end up going too far. Knowing what the brand stands for is not enough; you also have to define what the brand should not do.

Of course, any significant departure from what people have come to expect from a brand will cause concern. A recent article in the *New York Times* provocatively asked, "Has Lego Sold Out?" and documented concerns over the brand's shift from simple plastic building bricks into the more contemporary domains of Lego-related video games, children's books, and a TV spinoff. Referring to the creation of a Lego "multi-media juggernaut," Matt Richtel and Jesse McKinley write: "But some parents and researchers worry that the company's gain has come at a cost to its tiny consumers: diminishing the demand for their imagination, the very element that made the Lego brand famous in the first place."[17]

But has Lego really sold out? That depends on what you believe makes Lego meaningfully different from other children's toy brands. Do the Lego bricks attract kids simply because children enjoy the act of building something, or is there more to it? Lego's website states that the 80-year-old brand aims to develop "children's creativity through playing and learning." Are the bricks, video games, and stories all encouraging kids to use their imagination? If Lego stands for "developing creative minds," then perhaps there is no conflict between the different product channels.

Ultimately, whether or not the brand can extend into new experiences rests with the kids who want to use the products and the parents who buy them. The fact that *The Lego Ideas Book*, which challenges kids to "Unlock Your Imagination," has been on the *New York Times* children's bestseller list for 32 weeks suggests the demand is there. Provided Lego continues to exploit and reinforce the brand's meaning, then how it does so matters far less than how well it does so.

◼◼◼ FIT YOUR EXTENSION TO YOUR MEANINGFUL DIFFERENCE

If you have managed to create a strong brand that resonates with consumers in one category and country, then the best way to promote financial value growth is to extend that brand as far as you can. While brand extension is by no means risk-free, it is less risky than launching a completely new brand. Whatever the mechanism for brand extension, I would propose that most brands would do well to ensure that the extension fits with what the brand stands for and what the business can deliver. Richard Branson may be unconcerned by failure, but most companies do not take kindly to write-offs.

Questions to Consider

1. Have you extended your brand into all the categories in which it could reasonably compete? Have you extended it into countries where you think it could have a purpose not already served by incumbents?
2. Have you identified the stretchability of the brand, looking at all the possible categories in which it might exist and working out the economic feasibility of making this extension happen at the price point that you need to charge? Look at the prospects for growth in the relevant categories and explore the investment needed to move into new countries. Look at the likely returns and the time periods of those returns and make sure that you understand the barriers in the markets, the level of competition, and the opportunity to highlight your meaningful difference in that market. Consider licensing as an alternative to direct investment.

Chapter 15

Make the Most of Your Brand

Fact: If you work in marketing, then you care a heck of a lot more about your brand than your consumers do.

Want some proof? Millward Brown uses a cognitive science technique known as emotional priming to assess the speed and nature of our instinctive response to brands. The magnitude of the effects of brands that we observe using standardized tests suggests that brands really don't matter to consumers as much as we might expect them to. Typically, consumers' instinctive response to brands is weak—neither overly positive nor overly negative. Brands evoke far stronger responses from the people that work with them. The average response of Coca-Cola board members when presented with the Coke brand was twice the highest consumer response recorded in all our testing.

But why might this be? For a start, marketers tend to think about their brand for most of their working hours. Their brand is really, really important to them. It is their livelihood. Even when using a brand, consumers devote very little conscious attention to the brand itself. It is a means to an end, not something to be considered in its own right. The brand with which you spend the most time is probably your mobile phone, particularly if it is a smartphone. But do you actually think about the brand or even the product when using your phone? Or do you think about what you are trying to achieve? I am willing to bet that your focus is on the task at hand—calling your mother, checking email, or playing Angry Birds. Once the honeymoon period is over, you take your phone for granted, and the only time you really think about it is when something goes wrong. If asked why you like your phone, you probably struggle to come up with more than a few general statements about it. And yet your phone is important, right? It is such a necessary part of your life that you take it for granted.

Now think about someone who does not use the same brand of phone. How much time and conscious attention does he or she devote to the brand of your phone? The answer is virtually none. So his or her impression of that brand is gained through a series of disconnected encounters with it: seeing you use it on the street, seeing an ad for it, or glancing at an online review. An impression is all that he or she will get. Think about your Facebook profile. What impression would someone have of you based on the posts and photos on your timeline? Probably a very fragmented and distorted one, if your timeline is anything like mine. If a brand is to stand any chance of being meaningful, then it needs to ensure that every encounter delivers a consistent impression.

BRANDS EXIST AS A GENERAL IMPRESSION

People respond to impressions of the world around them, not to specific attributes. We don't consciously remember everything we have ever experienced about a brand; rather, we retain a general impression of it.

Gordon Pincott uses the word "gist" to describe this mental impression. This impression is all-important, because whatever comes to mind first when we think about a brand will help to define our opinion of it or whether we even consider it at all. Gordon states: "We often make purchasing decisions in matters of seconds with little cognitive effort. These intuitive decisions don't weigh up features and benefits; they simply call upon our sense or gist of a brand and make a decision on that basis. So the first-to-mind thought about a brand is going to be crucial in whether it gets chosen or not."[1]

In order for the gist to have an effect, it must first be triggered by something. It could be the color, text, sound, shape, or smell that triggers our impression. Knowing what the most salient triggers are and the strength and attractiveness of those impressions is really important, because these triggers help determine what is featured in ads and in-store to create a more coherent impression of the brand across all aspects of the brand experience (for more on this topic, see Gordon Pincott's "Keys to Brand Success."[2]

WHAT WE CAN LEARN FROM ICONIC BRANDS

In today's complex and busy world, brand names are everywhere—plastered all over websites, inside subway cars, on the sides of buses, and even in public toilets. But most of the time, even though they're accepted as part of the scenery, these brand names don't signify much to those who observe them.

Yet among the mindless white noise of modern marketing, a few brands stand out from the crowd. Admired and packed with meaning, these truly iconic brands inspire passion and fierce loyalty among their customers.

They represent the gold standard of branding. Any marketer worth his or her salt would love to work with an iconic brand—or better yet, help create one.

In his book *How Brands Become Icons*, former Oxford University professor Douglas Holt sheds light on the nature and origins of iconic brands.[3] He asserts that iconic brands respond to a society's desires and cultural tensions by drawing on their own unique myths and stories. However, the anecdotal nature of Holt's approach makes his findings difficult to apply to less distinguished brands, and most brands cannot hope that a societal trend or need will suddenly make their brand meaningful and result in its adoption as a cultural icon. So what universal themes apply to *all* brands, not just to those that are already iconic? I believe that iconic brands are ones that people find powerfully meaningful, different, and salient. I would define an iconic brand as an immediately recognizable brand, one that embodies a widely understood meaning that transcends its category and resonates within a specific culture.

AN AMERICAN ICON

By this definition, the Oreo cookie, which turned 100 years old in 2012, is iconic in the United States. The brand is immediately recognizable and tangibly different—the dark embossed biscuits sandwiching the white cream center make Oreos easily distinguishable from any other cookie. In fact, the Oreo is so recognizable that it can't be mistaken for anything else and can even be used as a reference point. Once, while driving in Maine, I was directed to look for a turn by watching for a field full of "Oreo cows." Sure enough, the black and white Belted Galloways made a good navigation aid. And Oreos do not just look different; they taste different too.

Oreos are also meaningful. In the United States, the Oreo stands out in people's minds for the sensory rituals they associate with eating it (twisting the wafers apart, licking the cream, dunking the wafers in milk) and the warm feeling of sharing those experiences with family and friends. Those good feelings are kept alive with ads that show people interacting over Oreos: brother with brother, mother with son, grandfather with granddaughter. These idyllic depictions represent family life as we might wish to experience it, and they evoke a powerful emotional response. As a result, the brand and its implicit associations of sharing and connection remain salient in peoples' minds.

Oreos are immediately recognizable and embody a widely understood meaning that resonates within American popular culture, but the understanding is not necessarily shared elsewhere in the world. For instance, for the brand to become successful in China, it was necessary to adapt the product and introduce consumers to the experience of dunking.

Twist, lick, and dunk is not a familiar ritual in China, and if you don't grow up with it, the roasted bitter taste of the cookie is unexpected. In order to make the brand successful, Kraft not only changed the cookie recipe but also played around with the cookie's shape and added new cream colors and flavors like green tea. However, the marketing team decided that without the experience of dunking the cookie, the brand was not going to reach its full potential. So a series of TV commercials was created in which cute kids introduced adults to the fine art of dunking an Oreo. The idea took off, and this meaningfully different experience helped Oreos become the bestselling cookie in China.[4]

An iconic brand is recognizable not just because it has invested in advertising but because it delivers a powerful brand experience that is founded in the brand's purpose. Oreos would not be iconic in the United States if people didn't think they tasted good. As Holt proposes, the meaning attached to iconic brands transcends functional benefits, but that does not mean that the brands have ignored those benefits (or that they can afford to ignore them in the future). Therefore, in terms of evaluating a brand's iconic potential, I first look at the brand's ability to meet a specific functional need. A brand that can meet a need or gratify a desire in a unique and meaningful way has an opportunity to build the strong emotional attachment that is the cornerstone of iconicity.

▬ THE "PEOPLE'S CAR" THAT BECAME AN ICON

In 1933, Ferdinand Porsche was charged by German chancellor Adolf Hitler to develop a car for the masses. Critical to the success of this mission was that the Volkswagen (literally, "people's car" in German) have the ability to transport two adults and three children at 100 km/h (62 mph) and be sold at a price point that would make it affordable for average working people.

First produced in 1938, the Volkswagen T1, later nicknamed the Beetle, became one of the most iconic vehicles of all time. Some denigrated the unique design, but it was certainly different, and after World War II, the reliable, economical, and affordable vehicle was exactly what the impoverished people of Germany required. Those same characteristics later appealed to hippies and others who were pursuing alternative lifestyles during the 1960s. The Beetle appeared in many movies and spawned a number of other nicknames, a sure sign of a brand that is vital and has shared cultural meaning. The 1998 reintroduction of the Beetle was successful in large part because it tapped into those positive feelings while bringing the product up to date.

In combination with its quirky looks, the Beetle's ability to meet people's transportation needs kindled the public affection that ultimately made the car an icon. Look behind the symbolism of most iconic brands, and you will find someone with a vision of how a product could serve a specific need better than the existing alternatives. When a brand's purpose or ideal resonates with a particular group of people, the brand moves one step closer to becoming iconic.

For instance, Google originated as a research project by Larry Page and Sergey Brin, an effort to find a better way to rank the relevance of search results than simply counting the appearances of a search term on a page. By redefining the way people used the Web, Page and Brin addressed the need to organize digital information and make it both widely accessible and useful. The intersection of the widespread need to find information quickly and the better solution offered by Google, which included a simple and uncluttered user interface, resulted in Google becoming the success it is today.

■■■ MEANING THAT IS SYMBOLIC AND SALIENT

One of Douglas Holt's iconic brand principles is that iconic brands develop identity myths that address people's desires and anxieties. The Marlboro man represents the values of the Western frontier: he is strong, independent, and capable. By presenting a consistent image over time, the brand has come to embody the frontier myth that now proves particularly appealing in developing economies. Brands like Mercedes, Rolex, and Montblanc have become symbolic of status and success. Jack Daniel's Tennessee Whiskey is symbolic of independence and has been an integral part of pop culture since the days of Frank Sinatra. The iPhone has come to symbolize coolness and hipness, and everyone knows that diamonds are forever. A glimpse of the uncluttered Google homepage, the Beetle's distinctive silhouette, or the black and white Oreo will instantly evoke recognition and trigger an intuitive understanding of the brand.

■■■ FIVE PRINCIPLES FOR THE EVERYDAY BRAND

Reaching iconic status is the Everest of the marketing world, and the vast majority of brands won't reach the summit. That said, continually referencing and seeking to improve your brand's performance using the following principles will help drive financial value growth.

1. (Re)discover and Stay True to Your Brand's Purpose

Many brands are founded and built around a specific purpose or ideal that subsequently fades from sight as leaders lose track of what originally made the company special. In these cases, restoring the original focus can often turn a struggling brand around. Sometimes, however, a company's purpose has to change because the world around it has changed. For example, twenty years ago, IBM shifted its purpose from making computers and chips to building a smarter and more efficient planet. This sort of transformation requires a total commitment, as Louis Gerstner details in his book *Who Says Elephants Can't Dance?*[5] This commitment includes the willingness to make changes that align structure with strategy and ongoing reinforcement of the new purpose in both internal and external communications.

2. Critically Examine the Experience Your Brand Delivers

You can have the noblest purpose in the world and still fail to deliver on people's expectations. For example, environmentally responsible ("green") brands want to ensure a better future for our planet, but all too often, their functional performance fails to satisfy consumers, or their price is so high that trial is inhibited. One brand, however, stands out from other green brands by not only meeting but also exceeding expectations. The household cleaning brand Method packs a one-two punch: its nontoxic and sustainable products are just as effective as traditional cleaners, and Method's unique design ethos enhances its products with packaging that is both distinctive and beautiful. These qualities could help Method become an iconic brand.

3. Make Sure Your Brand Resonates

The sweet spot for any brand is found at the intersection of what the brand stands for and the motivations and desires of its target audience. It is the elusive element that makes sure the brand resonates with its consumers. At the beginning of this book, I referred to the need to gain an intimate understanding of a brand's audience—for the marketer to put him or herself in the consumers' shoes. Look back through all the examples of brands that have created significant value growth. All of them rely on some insight into the target consumers' unspoken needs or desires and the creativity to meet those needs in a way that does not just satisfy but also enthuses and inspires. American Express OPEN, Blue Moon, and Coca-Cola have all found ways to connect with people and evoke a positive response.

4. Make a Difference

If there is one thing that you should take away from this book it is that you should be seeking to differentiate your brand—not to be different for difference's sake, but to provide people a reason to choose your brand and justify a price premium.

Importantly, that differentiation needs to stem from the brand itself, not the competition. Yes, you need to know those brands against which you are competing and differentiate your brand from them, but don't let the competition define your brand. Don't look at your competitors' strengths and weaknesses and then identify something that makes your brand different from them. Start with your brand, its origins and experience, and then seek to make what it has to offer different from that offered by the competition. If you must respond to competitive initiatives, then make sure your response is consistent with your brand's meaningfully different experience. Don't follow their lead.

5. Amplify across All Touch Points

If you have identified what makes your brand meaningfully different from the competition, then effective implementation becomes so much easier. A brief statement of that meaningful difference can become the compass bearing for all actions taken on behalf of the brand. Is the action under consideration consistent with that definition? Yes? Then do it. No? Don't do it. Work to build that consistent impression of the brand across every point of contact between the consumer and your brand. That way, you will build up a strong, coherent, and salient image of what the brand stands for in people's minds.

◾◾◾ FIVE THINGS YOU SHOULD DO NOW

The principles of creating a strong, valuable, and growing brand are pretty simple, but to put them into practice requires discipline and focus. The biggest challenge is ensuring that your colleagues share the same belief that you do. That means you need to be crystal clear about the opportunity for growth and what it is worth and able to communicate that idea convincingly to others.

1. Create Alignment within Your Company

Make sure that the understanding of what makes your brand meaningfully different is shared across all the brand's stakeholders.

In *Grow*, Stengel details how the Pampers diaper brand was revitalized as a result of alignment behind the brand ideal of helping moms raise happy, healthy babies.[6] The brand had been struggling in the late 1990s in the face of increased competition from Kimberly-Clark's Huggies and Pull-Ups brands. Stengel recalls how the Pampers brand benefited when the leadership team started hiring people who were passionate about babies and wanted to improve the lives of both moms and babies through their work.

However, Stengel also highlights how early communication of the brand ideal went astray, even after the team had identified its new brand ideal and started hiring the right people. In spite of the team's enthusiasm and positive feedback from research, the first iteration of marketing communication that focused on the ideal of helping moms raise happy, healthy babies did not resonate with consumers. Not everyone in the organization or its ad agencies understood and accepted the wider vision of the brand's relationship with moms and babies. As a result, early advertising overpromised how a diaper would help a baby's development. Stengel reports one mom commenting, "A diaper is not going to send my baby to Harvard." Other ads focused on a baby's development but failed to demonstrate how the brand would help in that development.

A lack of understanding about what makes a brand meaningfully different is a common problem. Even if the core marketing team has identified a clear and compelling positioning—one that is meaningful and different—many of the people who need to implement that positioning through the brand experience and communications will not understand it in the same way. It helps if people working on the brand have a personal knowledge of and interest in the product category, but that is not everything. It is very easy to misconstrue what someone else says and interpret it in the light of our own experience, and our experience may be idiosyncratic and not shared by most of the brand's consumers.

In many companies, even the people working on a brand will have little personal understanding of the category. That was part of Pampers' problem in the late 1990s. As Stengel notes, the brand made little effort to recruit people who wanted to work on products for babies. The organization was dominated by what the engineers thought would improve the product's absorbency and dryness—not by a concern with how the babies and their moms would feel about the product. How much more difficult is it, then, for people working in the brand's agencies to really understand and feel what the brand stands for? Several times, after conducting a ValueDrivers workshop that included members of the brand's ad agency, these individuals have stated, off the record, that they have a far better understanding of what the brand they are promoting stands for than they did before the workshop. One of the most important jobs in marketing today is to ensure a consistent understanding of the brand's meaningfully different experience across all business units and the brand's agencies.

2. Configure the Brand Experience around Your Brand's Meaningful Difference

Make sure that your brand experience is as positive and meaningful as possible. If you are responsible for all aspects of the brand experience, then make sure that the brief for product innovation and service development is designed to enhance and reinforce the meaningful difference. If you are not responsible for the development of your brand's product or service, then you need to get involved. Find ways to influence that process and make sure the consumer's needs come first and are not treated as an afterthought. Take a lesson out of Apple's playbook and design the brand experience to be simple and easy. Don't make people think; help them navigate your brand with as little conscious thought as possible. Make sure that the brand experience is aligned with what will make the brand meaningfully different and then innovate around that idea to build credibility.

Work to make sure your brand has a strong visual identity that is instantly recognizable—but don't stop there. A brand with powerful cues across all the senses has an intrinsic advantage over others. Identify symbols associated with the brand that add meaning to it and work to ensure that these cues evoke positive and motivating brand associations. Provided the same recognition cues are featured in both broadcast and in-store communications, they will allow the brand to realize synergies across marketing and sales channels.

3. Balance the Authentic with the Contemporary

Unlike iconic brands of days gone by, the Oreo brand is still a vital one. It is alive and kicking, doing things that make sure it is still salient and accessible. As John Ghingo, senior director for global biscuits at Kraft Foods, states in the *New York Times*: "What we have really set out to do is something authentic for the brand, true to its roots, that is fresh for today....We want to acknowledge what Oreo has been over the past 100 years and recognize what it means in today's world."[7]

That aim speaks to the fact that for an iconic brand to remain relevant, it must somehow remain true to its origins, but it cannot do that at the expense of its contemporary relevance. It needs to balance the two. It needs to remain vital.

One of the biggest challenges for any brand is to stay contemporary without unnecessarily changing what the brand stands for. The brands that do this successfully manage to apply the stories and values of their heritage to contemporary circumstances. For example, Jack Daniel's features its heritage in its advertisements with scenes of the whiskey being made, and company-sponsored concerts that bring people together still reflect the spirit and backcountry myths of the brand.

4. Don't Let Go

Over the last decade, there has been a lot of talk about the need for marketers to "let go," to cede control of their brand to its consumers. With the advent of the Internet and social media, so the theory goes, marketers can no longer dictate what people think or say about their brands. Brands now belong to their consumers.

The truth is that marketers have never controlled what people think and say about their brands, but marketers should control what the brand does and says. Strong brands lead. They shape their interactions and conversations with consumers—they do not simply respond to them. The actions and words of the brand need to be designed to resonate with its target audience, and the way to do that is to create a meaningfully different experience, one that leaves people feeling empowered and with a sense of confidence and self-esteem unique to that brand. Once that experience is established, the marketer needs to amplify the meaningful difference in everything the brand does.

If there is any truth to the statement that marketers need to let go, then perhaps it lies in the idea that we need to move from hard marketing to soft. Hard marketing is the world of indoctrination, argument, and persuasion. Soft marketing is the world of ideals, appeal, and service. Rather than telling people how much they should appreciate your brand, demonstrate how good it feels to use it. Rather than trying to convince, intrigue. Earn loyalty with a desirable brand experience—don't buy it with price promotions.

5. Stay Focused

Faced with aggressive competition, fragmenting media, and ever-faster feedback, brand marketers can be overly reactive. It may seem that doing something—even the wrong thing—is better than doing nothing. But if you keep in mind what makes your brand meaningfully different from the competition, then you can avoid this trap. Knowing what is truly meaningful to your customers will help you choose the right actions. Red Bull focuses all of its events on the idea of uplifting mind and body, offering both a spectacle and an unparalleled experience that is true to the brand's purpose. Felix Baumgartner's 2012 Red Bull-sponsored supersonic free-fall from 23 miles above the earth was an incredible feat and entirely consistent with what makes Red Bull meaningful and different.

The flip side of knowing what to do is knowing what not to do. All too often, brands end up getting lazy or greedy and ruin the very thing that makes them valuable by undermining their meaningfully different experience. Even strong brands are vulnerable to thoughtless marketing ploys, cost cutting, and overextension. Defining what your brand is not and what it must never do may be just as important as defining what it should do.

CREATING A MEANINGFULLY DIFFERENT BRAND STACKS THE ODDS IN YOUR FAVOR

In a letter to Jean-Baptiste Leroy, Benjamin Franklin commented on the likely permanence of the U.S. Constitution, saying, "In this world nothing can be said to be certain, except death and taxes."[8]

This is an adage that marketers would do well to remember. Just as there are no certainties in life, there are no certainties in marketing. When marketing a brand, the best we can hope for is to maximize the potential of a positive outcome and minimize the risk of a negative one. We cannot ensure that someone sees our ad, follows our brand on Twitter, or ends up buying it; we can only create the conditions that maximize the probability that they will do so. We cannot guarantee success, if for no other reason than the fact that our actions do not take place in a vacuum. Not only must we identify a compelling strategy and execute it well, but we must also do so better than the competition, who will be working just as hard to undermine our efforts.

To maximize the probability that people will buy our brand, we need to start by identifying what makes the brand meaningful and different. Meaningful differentiation is most powerful when it is tangible and functionally based. The Toyota Prius, the Nintendo Wii, and Red Bull are all brands that created new categories by moving outside the established norms of their product category. By stepping outside the bounds of their categories, these brands created a space that they can now call their own. Their advantage may not last, but the Nissan Leaf, Microsoft's Kinect for Xbox 360, and Hansen's Monster Energy not only have to deliver a compelling product experience but must also overcome mental barriers to competition.

Being salient—first to mind when a need arises—has always been a powerful driver of sales, and being known as the category creator is a powerful way to gain that position.

Of course, for many brands, there is little opportunity for significant, tangible differentiation. The marketer then needs to create that differentiation, either by highlighting an aspect of the brand experience to best advantage or by creating an emotional halo around the brand—or, ideally, both. There are few brands that are so well known that they can assume people have an intimate understanding of the experience they offer. Even when the brand is well known, its emotional appeal needs to reflect the brand's purpose and history. Coca-Cola has successfully enlisted the universal emotion of happiness to power its marketing. While the emotion is generic, it is a good fit with Coca-Cola's product experience and heritage. Whatever the basis of their personal understanding of the brand—whether it is based on personal memories of sipping cold Coke on a hot summer's day or on the recollection of the brand's Christmas advertising—many people will find that the brand's association with shared happiness just feels right. And that is a valuable thing. In 2013, Coca-Cola ranked fifth in the BrandZ Top 100 Most Valuable Global Brands Ranking, with its brand valued at $78,415 million, up 89 percent since it was first measured in 2006.

GROW YOUR BRAND'S MEANING TO GROW ITS FINANCIAL VALUE

Though times have changed, the foundational principles of good marketing have not. People still value things that they find meaningful and are predisposed to choose things that stand out from the crowd. Strong, profitable brands are meaningful to their consumers, are perceived as different from the competition, and are more salient than their competitors—they come to mind more quickly and easily than the alternatives.

Brands exist in consumers' minds as a network of associations and feelings. Marketing should seek to shape, enhance, and strengthen motivating associations that will lead to financially valuable behavior—a predisposition to buy the brand and pay the price asked and a willingness to buy it again. To do that, however, marketers must see the world as their consumers do, not through the lens of personal objectives and experience. Market research of all kinds will help inform that viewpoint, but only if marketers are willing to engage with that research in order to identify those opportunities that will make their brand more meaningful and valuable to its consumers.

Importantly, however, the understanding of the brand's meaning must be shared across all corporate and agency stakeholders in order to create an experience that exemplifies its purpose. A clear, succinct statement of what makes the brand meaningfully different in the minds of its consumers is the standard against which all actions should be judged. Does an action enhance that meaning or dilute it? Failure to align the brand experience with expectations created in marketing communications will only undermine the long-term value of the brand as disenchanted consumers tweet their discontent and walk away. Successful alignment will result in satisfied consumers who are more than willing to advocate a great brand experience and pay for it.

I'd love to end this book with one of those grandiose claims such as "Follow the principles outlined in this book and your brand will grow its financial value in 30 days or less!" I can't. I am certain, however, that the thoughtful application of the ValueDrivers framework will create the opportunity to boost your brand's sales and profits over the long term. That is what great marketing does. Marketing creates the opportunity to boost sales and profits over and above what a business could generate anyway and does so consistently over time.

Notes

Preface

1. Bill Lee, "Marketing Is Dead," *HBR Blog Network* (blog), *Harvard Business Review*, http://blogs.hbr.org/cs/2012/08/marketing_is_dead.html?awid=82663748019147 84340-3271. (Last accessed April 20, 2013.)

Part I

1. For the sake of consistency, I have used U.S. dollars, converted at the current exchange rate, throughout the book.
2. "Cadbury India records 40% growth on aggressive marketing drive," *Economic Times*. http://articles.economictimes.indiatimes.com/2012–01-04/news/30588888_1 _managing-director-anand-kripalu-cadbury-india-cadbury-plc. (Last accessed April 20, 2013.)

Chapter 1

1. Nigel Hollis, *The Global Brand* (New York: Palgrave Macmillan, 2008).
2. Dan Pallotta, "A Logo Is Not a Brand" *HBR Blog Network* (blog), *Harvard Business Review*, http://blogs.hbr.org/pallotta/2011/06/a-logo-is-not-a-brand.html. (Last accessed April 20, 2013.)
3. In this section I have used the word *customer* to be consistent with Pallota's use of the word. In the rest of the book I will use *customer* when referring to people buying a service and *consumer* when referring to people buying a product. This distinction acknowledges that many companies sell their product to retailers—their customers—who then sell the brand onto the end consumer. If the reference is generic, I will default to using *consumer*.
4. Roy Sutherland, "What Is Value?" YouTube video, 25:13, *TedxTalks*, posted October 22, 2010, http://youtu.be/3CjjHsE133w. (Last accessed April 24, 2013.)
5. Les Binet and Peter Fields, "Thinkbox Live Stream: Advertising Effectiveness," 3:26:32, from a Thinkbox presentation held December 5, 2012, posted by Brand Republic, December 5, 2012, 11 minutes and 29 seconds. http://www.brandrepublic .com/bulletin/brandrepublicnewsbulletin/article/1162684/thinkbox-live-stream -advertising-effectiveness/?DCMP=EMC-CONBrandRepublicdailynewsbulletin (Last accessed April 20, 2013.)
6. Mark Murray, interview with author, August 10, 2012.

7. Steve Mustardé, "From Whisky Producer to Global Icon: The Story of 'Keep Walking'" (BBH, IPA Effectiveness Awards case study 2008) © Copyright IPA, Institute of Practitioners in Advertising, London 2008. Accessed via Warc.com © 2013. Copyright and database rights owned by Warc.

8. Lonnie Miller, "Americans Are Holding Their Vehicles Longer: Is it Good for Loyalty?" *Polk Blog*, December 17, 2012, http://blog.polk.com/blog/blog-posts-by -lonnie-miller/americans-are-holding-their-vehicles-longeris-it-good-for-loyalty. (Last accessed April 24, 2013.)

9. "International Comparisons: The Handset Replacement Cycle" Mobile Future, http://mobilefuture.org/resources/international-comparisons-the-handset-replacement -cycle-2/. (Last accessed April 20, 2013.)

10. Leigh Van Valen, "A New Evolutionary Law" in *Evolutionary Theory* (1973), 1–30.

11. Economist Intelligence Unit, "Outside Looking In: The CMO Struggles to Get in Sync with the C-Suite" (Economist Intelligence Unit Ltd., 2012), www.management thinking.eiu.com/sites/default/files/downloads/EIU_SAS_CMO_WEBr1.pdf. (Last accessed April 24, 2013.)

12. Forrester Research and Heidrick & Struggles, "The Evolved CMO, 2012," 2012, http://www.heidrick.com/PublicationsReports/PublicationsReports/HS_Evolved CMO2012.pdf. (Last accessed April 24, 2013.)

Chapter 2

1. Katherine Harmon, "Motivated Multitasking: How the Brain Keeps Tabs on Two Tasks at Once," *Scientific American.*, April 15, 2010, http://www.scientificamerican .com/article.cfm?id=multitasking-two-tasks. (Last accessed April 24, 2013.)

2. Brian McElree, "Working Memory and Focal Attention," *Journal of Experimental Psychology: Learning, Memory and Cognition* 27, no. 1 (2001): 817–35.

3. Sadie F. Dingfelder, "A Workout for Working Memory," *Monitor on Psychology* 36, no. 8 (2005): 48, http://www.apa.org/monitor/sep05/workout.aspx. (Last accessed April 24, 2013.)

4. Daniel Kahneman, *Thinking, Fast and Slow* (New York: Farrar, Straus and Giroux, 2011).

5. "First Impressions Count on the Web," BBC News, January 16, 2006. http://news .bbc.co.uk/2/hi/technology/4616700.stm (Last accessed April 24, 2013.)

6. Gerd Gigerenzer, *Gut Feelings: The Intelligence of the Unconscious* (New York: Penguin Group, 2007).

7. Jack Neff, "Pick a Product: 40% of Public Decide in Store," *Advertising Age*, July 28, 2008, http://adage.com/article/news/pick-a-product-40-public-decide-store/129924/. (Last accessed April 24, 2013.)

8. Les Binet and Peter Fields, "Thinkbox Live Stream: Advertising Effectiveness," 3:26:32, from a Thinkbox presentation held December 5, 2012, posted by Brand Republic, December 5, 2012, 1 hour, 7 minutes and 43 seconds. http://www.brand republic.com/bulletin/brandrepublicnewsbulletin/article/1162684/thinkbox-live -stream-advertising-effectiveness/?DCMP=EMC-CONBrandRepublicdailynews bulletin. (Last accessed April 20, 2013.)

9. Nick Chater, on human decision making (presentation, Unilever Masterclass, Tenafly, New Jersey, October 25, 2011).

10. Peter Ayton "The Rationality of Human Judgment and Decisions," presentation, Unilever Masterclass, Fort Lee, New Jersey, November 29, 2012.

11. Graham Page, interview with author, January 22, 2013.

12. Gordon Brown, "Big Stable Brands, and Ad Effects: Fresh Thoughts about Why, Perhaps, Consistent Promotion Keeps Them Big," Admap, May 1991, © Warc 1991 http://www.warc.com/Content/ContentViewer.aspx?MasterContentRef=46945bb8 -caf1-4820-9a7b-d3ee9495ec4d.

13. Ibid.

14. Ibid.

15. Andy Farr and Gordon Brown, "Persuasion or Enhancement? An Experiment," MRS Conference Proceedings (1994): 69–77.

16. Byron Sharp, How Brands Grow: What Marketers Don't Know (Australia: Oxford University Press, 2010).

17. Ibid., 112.

18. Maurice Saatchi, "The Strange Death of Modern Advertising," Financial Times, June 22, 2006. http://www.ft.com/intl/cms/s/0/abd93fe6-018a-11db-af16-0000779e2340 .html. (Last accessed April 24, 2013.)

19. Ayton, "Human Judgement and Decisions."

Chapter 3

1. Nigel Hollis, The Global Brand: How to Create and Develop Lasting Brand Value in the World Market, (New York: Palgrave Macmillan, 2008), 13.

2. Alex Pentland, interview with Robin Young, Here & Now, 90.9WBUR, NPR (Boston), May 26, 2011, http://hereandnow.wbur.org/2011/05/26/cell-phone-data. (Last accessed April 24, 2013.)

3. "Bill Moggridge—Clip from Objectified," YouTube video, 3:16, posted by "gary hustwit," August 29, 2012, http://vimeo.com/48488802 (Last accessed April 24, 2013.)

4. Hollis, Global Brand, 56.

5. Tim Broadbent, "BMW: How 15 Years of Consistent Advertising Helped BMW Treble Sales Without Losing Prestige," IPA Effectiveness Awards case study, 1994. © Copyright IPA, Institute of Practitioners in Advertising, London 1994. Accessed via Warc.com. © 2013 Copyright and database rights owned by Warc.

6. Barry Schwartz, The Paradox of Choice: Why More Is Less (New York: Harper Perennial, 2004).

7. Mark Batey, Brand Meaning (New York: Routledge/Taylor and Francis Group, 2008).

8. "Procter & Gamble's Tide Pods: The Future of Laundry Care or Expensive Folly?" Euromonitor International, April 10, 2012.

9. Jack Neff, "Tide Pods Winning $7 Billion Detergent Wars by Redefining Value," Advertising Age, December 18, 2012, http://adage.com/article/news/tide-pods -winning-7-billion-detergent-wars-redefining/238779/. (Last accessed April 24, 2013.)

10. John Reed, "Skoda Shifts Up a Gear into the Fast Lane," *Financial Times*, March 4, 2012, http://www.ft.com/cms/s/0/332791e4-663b-11e1-979e-00144feabdc0.html#ixzz 2KQvRMWIe. (Last accessed April 24, 2013.)

Chapter 4

1. Analysis conducted by Peter Field and Les Binet using the IPA Effectiveness Awards, by Nielsen Analytic Consulting using consumer packaged goods data, and by Millward Brown Mexico using client-supplied sales data; Les Binet and Peter Field, *Marketing in the Era of Accountability*. (London: Warc, 2007); Peter Field, "Account Planners Need to Care More About Share of Voice." *Admap*, September 2009, Issue 508.

Chapter 5

1. Mark Murray, interview with author, August 10, 2012.
2. Hans Kundnani, "Dyson Cleans Up," *Guardian*, November 1, 2006, http://www .guardian.co.uk/business/2006/nov/01/japan.money. (Last accessed April 24, 2013.)
3. Lee Ji-Yoon, "Woongjin Coway Enjoys Boon from National Health Trend," *Korea Herald*, October 7, 2011, http://nwww.koreaherald.com/view.php?ud=201107 10000224. (Last accessed April 24, 2013.)
4. Joon-Kee Hong, "CEO Greeting," Coway website, http://www.coway.com/about /greeting.asp. (Last accessed April 24, 2013.)
5. Jim Stengel, *Grow: How Ideals Power Growth and Profit at the World's Greatest Companies* (New York: Crown Business, 2011).
6. Ibid, 19.
7. Luciana Hashiba, "Story: Innovation in Well-Being—The Creation of Sustainable Value at Natura," May 18, 2012, http://www.managementexchange.com/story /innovation-in-well-being. (Last accessed April 24, 2013.)
8. José Vicente Marino, "Marca Natura" (presentation, launch event of the Top 50 Most Valuable Latin American Brands, June 21, 2012).
9. Clayton M. Christensen and Michael E. Raynor, *The Innovator's Solution: Creating and Sustaining Successful Growth*. (Harvard Business School Publishing Corporation, 2003), 99.
10. "CEO Meg Whitman on the Future of Hewlett-Packard," interview with Kai Ryssdal, podcast, *Marketplace*, American Public Media, November 20, 2012, http://www .marketplace.org/topics/business/corner-office/ceo-meg-whitman-future-hewlett -packard. (Last accessed April 24, 2013.)
11. Ibid.
12. Ibid.
13. "Taking AOL Beyond 'You've Got Mail,'" interview with Kai Ryssdal, podcast, *Marketplace*, American Public Media, December 11, 2012, http://www.marketplace .org/topics/business/corner-office/taking-aol-beyond-youve-got-mail. (Last accessed April 24, 2013.)

Chapter 6

1. Martin Lindstrom, *Brand Sense: Sensory Secrets Behind the Stuff We Buy* (New York: Free Press, 2005).
2. Joann Peck and Suzanne B. Shu, "The Effect of Mere Touch on Perceived Ownership," *Journal of Consumer Research* 36, no. 3 (2009): 434–47.
3. Ibid, 445.
4. Lindstrom, *Brand Sense*, 152.
5. Chris Clark, "Please Mind the Gap" presentation given at IAA Global Brand Think Tank, Four Seasons Hotel, Hampshire, UK, June 24–25, 2008.
6. Sarah Ostman, "Procter & Gamble Design Chief Discusses Product Design, Marketing in the Digital Age," Northwestern University, McCormick School of Engineering, October 18, 2011, http://www.mccormick.northwestern.edu/news/articles/article _989.html. (Last accessed April 24, 2013).
7. Jack Neff, "P&G Pushes Design in Brand-Building Strategy," *AdAge*, April 12, 2010, http://adage.com/article?article_id=143211. (Last accessed April 24, 2013.)
8. "Independent Lens—Objectified—Film Clip #1," YouTube video, 4:12, from a weekly series premiered on PBS November 24, 2009, posted by "IndependentLens," September 30, 2009, http://youtube/J_9n0SIxO6c. (Last accessed April 24, 2013.)
9. Nichola Groom, "Ford to Upgrade, Extend Warranty on Touch-Screen System," Reuters, November 28, 2012, http://www.reuters.com/article/2012/11/28/auto show-ford-infotainment-idUSL1E8MSAT520121128. (Last accessed April 24, 2013.)

Chapter 7

1. Jeremy Diamond, email exchange with author, February 8, 2013.
2. Ibid.
3. Dan Danner, "$5.5 Billion Spent on Small Business Saturday," *Forbes.com*, November 27, 2012, http://www.forbes.com/sites/groupthink/2012/11/27/5-5-billion-was-spent -at-small-businesses-on-small-business-saturday/. (Last accessed April 24, 2013.)
4. Sean Gregory, "Domino's New Recipe: (Brutal) Truth in Advertising," *Time*, May 5, 2011, http://www.time.com/time/business/article/0,8599,2069766,00.html#ixzz2Eg1y ITMM. (Last accessed April 24, 2013.)
5. Lisa Baertlein, "Olympics Boost Domino's Pizza Results; Shares Jump," Reuters, October 16, 2012, http://www.reuters.com/article/2012/10/16/us-dominos-results -idUSBRE89F0LS20121016. (Last accessed April 24, 2013.)
6. Domino's Pizza: Pizza Turnaround, ARF Ogilvy Awards: Grand Ogilvy Award, Retail and E-Tail case study, 2011. © Copyright Advertising Research Foundation 2011. Accessed via Warc.com. © 2013 Copyright and database rights owned by Warc.
7. Anna-Louise Jackson and Anthony Feld, "Domino's 'Brutally Honest' Ads Attract Sales as Consumer Spending Falters," Bloomberg, October 17, 2011, http://www .bloomberg.com/news/2011-10-17/domino-s-brutally-honest-ads-offset-slowing -consumer-spending.html. (Last accessed April 24, 2013.)
8. Gregory, "Domino's New Recipe."
9. Candice Choi, "Campbell Soup Tries to Reinvent Itself," *Huffington Post*, September 7, 2012, http://www.huffingtonpost.com/2012/09/07/campbell-soup-reinvents_n _1864170.html. (Last accessed April 24, 2013.)

10. Ibid.,
11. "Millennial Generation Soup Campaign," *The Colbert Report*, first broadcast November 15, 2012 by Comedy Central, http://www.colbertnation.com/the-colbert-report-videos /421325/november-15-2012/millennial-generation-soup-campaign. (Last accessed April 24, 2013.)
12. Ibid.
13. Ibid.
14. Comment made on Millward Brown's proprietary Greenhouse social network.

Chapter 8

1. Chris Zook and James Allen, "The Great Repeatable Business Model," *The Magazine* (blog), *Harvard Business Review* (November 2011), http://hbr.org/2011/11/the-great -repeatable-business-model/ar/1. (Last accessed April 24, 2013.)
2. Joan Magretta, "Stop Competing to Be the Best," HBR Blog Network, *Harvard Business Review*, November 30, 2011, http://blogs.hbr.org/cs/2011/11/stop_competing _to_be_the_best.html. (Last accessed April 24, 2013.)
3. Euromonitor International global briefing "From Traditional To Postmodern: The Evolution of Soft Drink Drinking Habits," August 2012.
4. Leander Kahney, "Straight Dope on the iPod's Birth," *Wired*, October 17, 2006, http://www.wired.com/gadgets/mac/commentary/cultofmac/2006/10/71956. (Last accessed April 24, 2013.)
5. Rob Walker, "The Guts of a New Machine," *New York Times*, November 30, 2003, http:// www.nytimes.com/2003/11/30/magazine/30IPOD.html?pagewanted=all&smid =pl-share. (Last accessed April 24, 2013.)
6. Ibid.
7. Justin Scheck, "Dell Tests Player to Renew iPod Battle," *Wall Street Journal*, July 30, 2008, http://online.wsj.com/article/SB121738346889295815.html. (Last accessed April 24, 2013.)
8. Sarah Everts, "How Advertisers Convinced Americans They Smelled Bad," *Smithsonian*, August 3, 2012, http://www.smithsonianmag.com/history-archaeology/How -Advertisers-Convinced-Americans-They-Smelled-Bad-164779646.html#ixzz2JC5ssWp7. (Last accessed April 24, 2013.)
9. "3-Minute AdAge," video, 2:58, *AdAge*, November 20, 2008, http://link.bright cove .com/services/player/bcpid1543292789?bctid=2708128001. (Last accessed April 24, 2013.)
10. Elaine Wong, "Concentrating on Bleach," *Adweek*, January 10, 2009, http://www .adweek.com/news/advertising-branding/concentrating-bleach-105074. (Last accessed April 24, 2013.)
11. Euromonitor International Category Briefing, "Bleach in the U.S.," May 2012.

Chapter 9

1. Bobbie Stein, "Reuben Mattus Scooped the Competition with His Pricey and Nonsense-Named Haagen-Dazs," *People*, August 17, 1981, http://www.people .com/people/archive/article/0,,20080001,00.html. (Last accessed April 24, 2013.)

2. Ibid.

3. Ibid.

4. Ibid.

5. Ibid.

6. General Mills, *A Portfolio for Global Growth: Annual Report 2011*, http://generalmills .com/~/media/Files/2011_annual_report. (Last accessed April 24, 2013.)

7. Leo Hopf and William Welter, *Rethink, Reinvent, Reposition: 12 Strategies to Renew Your Business and Boost Your Bottom Line* (Avon, MA: Adams Business, 2010), 127.

8. Gary Chu, interview with the author, November 13, 2012.

9. David Kaplan, "General Mills' Global Sweet Spot," *Fortune*, May 16 2011 http:// management.fortune.cnn.com/2011/05/16/general-mills-global-sweet-spot/. (Last accessed April 24, 2013.)

10. Ibid.

Part III

1. Tom Buday, interview with author, September 28, 2012.

Chapter 10

1. Craig Twyford, "What Makes Big Brands Stay Big?" Nielsen Featured Insights, March 2010, http://es.nielsen.com/news/documents/BigBrandsFI.pdf. (Last accessed April 25, 2013.)

2. Ibid.

3. Harish Manwani, "Winning in the Market Place," Unilever, November 2012, http:// www.unilever.com/images/ir_Harish-Manwani-and-Alex-von-Behr-Winning-in -the-market-place_tcm13-330937.pdf. (Last accessed April 25, 2013.)

4. Rupal Parekh, "South Korea's Cheil Captures Media Grand Prix at Cannes," *Advertising Age*, June 21, 2011, http://adage.com/article/special-report-cannes /south-korea-s-cheil-captures-media-grand-prix-cannes/228319/. (Last accessed April 25, 2013.)

5. James B. Stewart, "Behind eBay's Comeback," *New York Times*, July 27, 2012, http:// www.nytimes.com/2012/07/28/business/ebays-turnaround-defies-convention -for-internet-companies.html?smid=pl-share. (Last accessed April 25, 2013.)

6. Kyle Spencer, "The Future of Facebook Is Mobile Payments," Seeking Alpha, July 30, 2012, http://seekingalpha.com/article/762711-the-future-of-facebook-is-mobile -payments. (Last accessed April 25, 2013.)

7. Stewart, "Behind eBay's Comeback."

8. Stephanie Clifford, "Once Proudly Web Only, Shopping Sites Hang Out Real Shingles," *New York Times*, December 18, 2012, http://www.nytimes.com/2012 /12/19/business/shopping-sites-open-brick-and-mortar-stores.html?smid=pl -share. (Last accessed April 25, 2013.)

9. Louise Lucas, "Unilever Extends 'Shakti' Scheme to Africa," *Financial Times*, August 30, 2011, http://www.ft.com/intl/cms/s/0/732067cc-d012-11e0-81e2-00144feabdc0 .html#axzz2FdIpd89A. 9. (Last accessed April 25, 2013.)

10. Hoag Levins, "Peter Arnell Explains Failed Tropicana Package Design," 3-Minute AdAge video, 2:58, *Advertising Age*, February 26, 2009, http://adage.com/article /video/peter-arnell-explains-failed-tropicana-package-design/134889/. (Last accessed April 25, 2013.)

11. Natalie Zmuda, "Tropicana Line's Sales Plunge 20% Post-Rebranding," *Advertising Age*, April 2, 2009, http://adage.com/article/news/tropicana-line-s-sales-plunge -20-post-rebranding/135735/. (Last accessed April 25, 2013.)

12. Natalie Zmuda, "Coke Takes Juice Lead from Pepsi," *Advertising Age*, February 1, 2010, http://adage.com/article/news/marketing-coke-takes-juice-lead-pepsi/141844/. (Last accessed April 25, 2013.)

13. Ibid.

14. Paul Marston, "BA Restores Union Flag Design to All Tailfins," *The Telegraph* (UK), May 11, 2001, http://www.telegraph.co.uk/news/uknews/1329843/BA-restores -Union-flag-design-to-all-tailfins.html. (Last accessed April 25, 2013.)

15. Ibid.

16. "2012 State of the Industry: Energy Drinks," *Beverage Industry*, July 18, 2012, http:// www.bevindustry.com/articles/85655-consumers-seek-out-energy-boosts. (Last accessed April 25, 2013.)

17. Ibid.

18. Barry Silverstein, "5-Hour Energy Powered Up," *Brandchannel*, January 13, 2010. http://www.brandchannel.com/features_profile.asp?pr_id=472. (Last accessed April 25, 2013.)

19. Clare O'Connor, "The Mystery Monk Making Billions with 5-Hour Energy," *Forbes*, February 8, 2012, http://www.forbes.com/sites/clareoconnor/2012/02/08/manoj -bhargava-the-mystery-monk-making-billions-with-5-hour-energy/. (Last accessed April 22, 2013.)

Chapter 11

1. Patrick Vlaskovits, "Henry Ford, Innovation, and That 'Faster Horse' Quote," *HBR Blog Network* (blog), *Harvard Business Review*, August 29, 2011, http://blogs.hbr.org /cs/2011/08/henry_ford_never_said_the_fast.html. (Last accessed April 25, 2013.)

2. Ibid.

3. "The Science Behind Watson," IBM, http://www-03.ibm.com/innovation/us /watson/science-behind_watson.shtml. (Last accessed April 25, 2013.)

4. *IBM: Watson case study*, Effie Worldwide: Gold, North America Effies 2012. Copyright Effie Worldwide, Inc. 2012. Accessed via Warc.com. © 2013 Copyright and database rights owned by Warc.

5. Ibid.

6. José Calderoni, interview with author, July 27, 2012.

7. Dede Fitch, "Celebrity Power: Can Less Be More?," Millward Brown, December 2006, http://millwardbrown.com/Libraries/MB_POV_Downloads/MillwardBrown _POV_CelebrityPower.sflb.ashx. (Last accessed April 25, 2013.)

8. "Lessons from the Jobsite," Cat, http://www.catfootwear.com/US/en/mikerowe WORKS. (Last accessed April 25, 2013.)

9. Normandy Madden, "Ugly Wudi Will Entertain Chinese Viewers through Summer 2010," May 13, 2009, *Advertising Age*, http://adage.com/china/article/china-news /ugly-wudi-will-entertain-chinese-viewers-through-summer-2010/136606/. (Last accessed April 25, 2013.)

10. Emily Bryson York, "Subway Can't Stop Jonesing for Jared," *Advertising Age*, February 18, 2008, http://adage.com/article/news/subway-stop-jonesing-jared /125142/. (Last accessed April 25, 2013.)

Chapter 12

1. Jonah Bloom, "JetBlue's Fares, Customer Experience Are Key to its Brand Building, Sales Success," *Advertising Age*, December 9, 2002. http://adage.com/article/news /jetblue-s-fares-customer-experience-key-brand-building-sales-success/36453/. (Last accessed April 25, 2013.)

2. Peter Field, "Account Planners Need to Care More about Share of Voice," *Admap* 508 (September 2009): 28–30.

3. Andrew Stanton, "The Clues to a Great Story," 19:16, from a TED Talk filmed February 2012, uploaded March 2012, http://www.ted.com/talks/andrew_stanton_the_clues _to_a_great_story.html. (Last accessed April 25, 2013.)

4. Ibid.

5. Antonio Damasio, *The Feeling of What Happens: Body and Emotion in the Making of Consciousness* (New York: Harcourt Inc., 1999).

6. Stanton, "Clues to a Great Story."

7. Ibid.

8. *Ocean Spray—Straight from the Bog*, Effie Worldwide: Gold case study, Renaissance, Effie Awards 2008 © Copyright Effie Worldwide, Inc. 2008. Accessed via Warc.com © 2013. Copyright and database rights owned by Warc.

9. Ibid.

10. Suzy Menkes, "Marketing to the Millennials," *New York Times*, March 2, 2010, http:// www.nytimes.com/2010/03/03/fashion/03iht-rmil.html. (Last accessed April 25, 2013.)

11. Caryn Rousseau, "Classic Fashion Brand Burberry Goes Digital," *Bloomberg Businessweek*, December 26, 2012, http://www.businessweek.com/ap/2012-12-26 /classic-fashion-brand-burberry-goes-digital. (Last accessed April 25, 2013.)

12. Richard L. Tso, " 'Digital Should Never Be an Afterthought': Burberry Ad Campaigns Blend Music, Fashion and Social," Adotas, January 2, 2013, http://www.adotas .com/2013/01/%E2%80%98digital-should-never-be-an-afterthought%E2%80%99 -burberry-ad-campaigns-blend-music-fashion-social/. (Last accessed April 25, 2013.)

13. Tom Buday, interview with author, September 28, 2012.

14. "Case Study: How Gangnam Style Went Viral with a Strategic Marketing Campaign from YG Entertainment," The Drum, October 30, 2012, http://www.thedrum.com /news/2012/10/30/case-study-how-gangnam-style-went-viral-campaign-yg -entertainment. (Last accessed April 25, 2013.)

15. Gordon Pincott, interview with author, January 29, 2013.

Chapter 13

1. Dominique M. Hanssens, ed., *Empirical Generalizations about Marketing Impact* (Cambridge, MA: Marketing Science Institute, 2009), 19.

2. Michael V. Marn, Eric V. Roegner, Craig C. Zawada, "The Power of Pricing," *McKinsey Quarterly* 1 (2003), http://download.mckinseyquarterly.com/popr03.pdf. (Last accessed January 30, 2013.)

3. Ibid, 29.

4. Michael Derepas and Mandy Henderson, "Bega Cheese: How Human Values Can Still Win against Price Discounting," The Communications Council: Gold, Australian Effie Awards case study, 2011. © Copyright The Communications Council 2011. Accessed via Warc.com © 2013. Copyright and database rights owned by Warc.

5. Diageo Half Year Results, presentation and script downloaded from 2012 half year results press kit. http://www.diageo.com/en-ie/newsmedia/Pages/resource.aspx?resourceid=1205. (Last accessed April 25, 2013.)

6. Sarah Nassauer, "A Hard-Working Suit," *The Wall Street Journal*, April 14, 2011. http://online.wsj.com/article/SB10001424052748703385404576258671135584478.html. (Last accessed April 25, 2013.)

Chapter 14

1. Harish Manwani, "Winning in the Market Place," Unilever, November 2012, http://www.unilever.com/images/ir_Harish-Manwani-and-Alex-von-Behr-Winning-in-the-market-place_tcm13-330937.pdf. (Last accessed April 25, 2013.)

2. "Unilever Facts," Unilever, http://www.unilever.com/aboutus/introductiontounilever/unileverataglance/index.aspx (last accessed April 25, 2013); Julian E. Barnes, "The Making (or Possible Breaking) of a Megabrand," *New York Times*, July 22, 2001, http://www.nytimes.com/2001/07/22/business/the-making-or-possible-breaking-of-a-megabrand.html?pagewanted=all&src=pm (last accessed April 25, 2013); "50 Years of Dove: The Story of a Brand (1955–2005)" Ogilvy & Mather Worldwide, http://www.mediabistro.com/portfolios/samples_files/9bJKVI0M6lMEDNp8z8hOlCQTt.pdf (last accessed April 25, 2013); "Dove Brand Profile," Dove, http://www.adbrands.net/files/us/dove_us_p.htm (last accessed April 25, 2013); Christine Bittar, "Dove 'Pro-Age' Will Woo Menopausal Women," *Marketing Daily*, January 16, 2007, http://www.mediapost.com/publications/article/53938/ (last accessed April 25, 2013); "Dove," http://www.unileverusa.com/brands-in-action/detail/Dove-/298217/ (last accessed April 25, 2013).

3. Nigel Hollis, *The Global Brand: How to Create and Develop Lasting Brand Value in the World Market* (New York: Palgrave Macmillan, 2008).

4. Gordon Pincott, interview with author, January 29, 2013.

5. Allen Hall, "First It Was Furniture, Then It Was Hotels…Now Ikea Wants to Build an Entire Town (But Will There Be Enough Screws Included?)," *Daily Mail Online*, August 22, 2012, http://www.dailymail.co.uk/news/article-2191967/Ikea-wants-build-entire-town-screws-included.html#ixzz25JQOdjgD. (Last accessed April 25, 2013.)

6. "Strand East," http://strandeast.com/#home. (Last accessed April 25, 2013.)

7. Louisa Schaeffer, "World's Biggest Retailer Wal-Mart Closes Up Shop in Germany," *DW*, July 28, 2006. http://www.dw.de/worlds-biggest-retailer-wal-mart-closes-up-shop-in-germany/a-2112746. (Last accessed April 25, 2013.)

8. Ellen Byron, "Mr. Clean Takes Car-Wash Gig," *Wall Street Journal*, February 5, 2009, http://online.wsj.com/article/SB123379252641549893.html. (Last accessed April 25, 2013.)

9. Howard Schultz, "The Commoditization of the Starbucks Experience" republished by *The Wall Street Journal*, February 24, 2007. http://online.wsj.com/public/article/SB117234084129218452-hpbDoP_cLbOUdcG_0y7qLlQ7Okg_20080224.html. (Last accessed April 23, 2013.)

10. David Teather, "Starbucks Legend Delivers Recovery by Thinking Smaller," *The Guardian* (UK), January 21, 2010, http://www.guardian.co.uk/business/2010/jan/21/starbucks-howard-schultz. (Last accessed April 25, 2013.)

11. Howard Schultz, "Looking Forward to Starbucks Next Chapter," *Starbucks Blog*, January 5, 2011, http://www.starbucks.com/blog/looking-forward-to-starbucks-next-chapter/643. (Last accessed April 25, 2013.)

12. Starbucks Corporation, *Fiscal 2011 Annual Report*, http://investor.starbucks.com/phoenix.zhtml?c=99518&p=irol-reportsannual. (Last accessed April 25, 2013.)

13. Teather, "Starbucks Legend Delivers."

14. Bob de Wit and Ron Meyer, *Strategy: Process, Content, Context* (Andover, UK: Cengage Learning EMEA, 2010), 319.

15. Sir Richard Branson, "The Collective Power Of a British Brand," a letter to the editor, *The New York Times*, published June 2, 2002. http://www.nytimes.com/2002/06/02/business/l-the-collective-power-of-a-british-brand-334391.html. (Last accessed April 23, 2013.)

16. Alan Deutschman, "The Gonzo Way of Branding," *Fast Company*, October 1, 2004, http://www.fastcompany.com/51052/gonzo-way-branding. (Last accessed April 25, 2013.)

17. Matt Richtel and Jesse McKinley, "Has Lego Sold Out?," *New York Times*, December 22, 2012, http://www.nytimes.com/2012/12/23/sunday-review/has-lego-sold-out.html?smid=pl-share. (Last accessed April 25, 2013.)

Chapter 15

1. Gordon Pincott, interview with author, January 29, 2013.

2. Gordon Pincott, "The Keys to Brand Success," October 2009. http://millwardbrown.com/Libraries/MB_POV_Downloads/MillwardBrown_POV_KeysToBrand Success.sflb.ashx. (Last accessed April 23, 2013.)

3. Douglas Holt, *How Brands Become Icons: The Principles of Cultural Branding* (Boston: Harvard Business School Publishing Corp, 2004).

4. Robert Smith, "Rethinking The Oreo For Chinese Consumers," *NPR Planet Money*, January 7, 2012. http://www.npr.org/blogs/money/2012/01/27/145918343/rethinking-the-oreo-for-chinese-consumers. (Last accessed April 25, 2013.)

5. Louis V. Gerstner, *Who Says Elephants Can't Dance?: Leading a Great Enterprise through Dramatic Change* (New York: HarperBusiness, 2002).

6. Jim Stengel, *Grow: How Ideals Power Growth and Profit at the World's Greatest Companies* (New York: Crown Business, 2011), Chapter 7.

7. Stuart Elliott, "The Oreo Turns 100, with a Nod to the Past," *New York Times*, February 27, 2012, http://www.nytimes.com/2012/02/28/business/media/the-oreo-turns -100-with-a-nod-to-the-past-advertising.html. (Last accessed April 25, 2013.)

8. Benjamin Franklin to Jean-Baptiste Leroy, November 13, 1789. *ThisDayinQuotes.com*, translated from the French, http://www.thisdayinquotes.com/2010/11/nothing-is -certain-except-death-and.html (Last accessed April 23, 2013.)

Index